Unconscious Networks

This book develops an original theoretical framework for understanding human–technology relations. The author's approach, which he calls technoanalysis, analyzes artificial intelligence based on Freudian psychoanalysis, biosemiotics, and Latour's actor–network theory.

How can we communicate with AI to determine shared values and objectives? And what, ultimately, do we want from machines? These are crucial questions in our world, where the influence of AI-based technologies is rapidly growing. Unconscious dynamics influence AI and digital technology, and understanding them is essential to better control AI systems. This book's unique methodology—which combines psychoanalysis, biosemiotics, and actor–network theory—reveals a radical reformulation of the problem of the human mind. Technoanalysis views the mind as a hybrid network of humans and nonhuman actants in constant interaction with one another. The author argues that human unconscious dynamics influence and shape technology, just as technology influences and shapes human unconscious dynamics. He proceeds to show how this conception of the relationship between the unconscious and technology can be applied to social robotics and AI.

Unconscious Networks will appeal to scholars and advanced students interested in philosophy of technology, philosophy of artificial intelligence, psychoanalysis, and science and technology studies.

Luca M. Possati is a researcher and lecturer at the Institute of Philosophy, University of Porto, Portugal. He is the author of many books, including *Software as Hermeneutics: A Philosophical and Historical Study* (2022) and *The Algorithmic Unconscious: How Psychoanalysis Helps in Understanding AI* (Routledge, 2021).

Routledge Studies in Contemporary Philosophy

Epistemic Injustice and the Philosophy of Recognition
Edited by Paul Giladi and Nicola McMillan

Evolutionary Debunking Arguments
Ethics, Philosophy of Religion, Philosophy of Mathematics, Metaphysics, and Epistemology
Edited by Diego E. Machuca

The Theory and Practice of Recognition
Edited by Onni Hirvonen and Heikki J. Koskinen

The Philosophy of Exemplarity
Singularity, Particularity, and Self-Reference
Jakub Mácha

Wealth and Power
Philosophical Perspectives
Michael Bennett, Huub Brouwer, and Rutger Claassen

Philosophical Perspectives on Memory and Imagination
Anja Berninger and Íngrid Vendrell Ferran

Unconscious Networks
Philosophy, Psychoanalysis, and Artificial Intelligence
Luca M. Possati

Updating the Interpretive Turn
New Arguments in Hermeneutics
Edited by Michiel Meijer

For more information about this series and a full list of titles, please visit: www.routledge.com/Routledge-Studies-in-Contemporary-Philosophy/book-series/SE0720

Unconscious Networks

Philosophy, Psychoanalysis, and Artificial Intelligence

Luca M. Possati

Routledge
Taylor & Francis Group
NEW YORK AND LONDON

First published 2023
by Routledge
605 Third Avenue, New York, NY 10158

and by Routledge
4 Park Square, Milton Park, Abingdon, Oxon, OX14 4RN

Routledge is an imprint of the Taylor & Francis Group, an informa business

ISBN: 978-1-032-38551-8 (hbk)
ISBN: 978-1-032-38552-5 (pbk)
ISBN: 978-1-003-34557-2 (ebk)

DOI: 10.4324/9781003345572

Typeset in Sabon
by Apex CoVantage, LLC

Contents

Figures

Tables

Acknowledgments

I thank the Institute of Philosophy of the University of Porto which made possible the development of this research. I also thank all the anonymous reviewers who allowed me to correct my mistakes. I express my gratitude especially to Andrew Weckenmann for accepting the volume in the Routledge Contemporary Philosophy series.

Introduction

How can we communicate with artificial intelligence (AI) to share our values and objectives? What are humanity's goals with respect to AI, and how can we clarify them? What do we want from machines? These essential questions in a world of increasingly pervasive AI systems are at the core of *Unconscious Networks*. This book intends to solve a crucial problem extensively addressed by AI researchers: the so-called problem of the "control" of AI (i.e., how we can transmit our desires, needs, and values to AI systems). This is an instance of what is known as the "alignment problem," whereby the "sorcerer's apprentice effect" is avoided; if we do not know how to convey our real needs and values to an AI system,

> we will find ourselves more and more often in the position of the 'sorcerer's apprentice': we conjure a force, autonomous but totally compliant, give it a set of instructions, then scramble like mad to stop it once we realize our instructions are imprecise or incomplete—lest we get, in some clever, horrible way, precisely what we asked for.
>
> (Christian 2020, 20)

How to prevent such a catastrophic divergence—or how to ensure that AI models capture our habits, norms, and values; understand what we mean or intend; and, above all, do what we want—"has emerged as one of the most central and most urgent scientific questions in the field of computer science" (Christian 2020, 20). As I will show in the following, the problem of the control of AI can be connected to the Collingridge dilemma (Collingridge 1980). This dilemma is "one of the biggest challenges for responsible design and innovation" (Kudina and Verbeek 2019, 1).

This book aims to develop a new approach to the problem of the control of AI and the Collingridge dilemma, assuming that they cannot be entirely solved technically and that they require the use of a different methodology. I follow a multi-method approach inspired by Freudian psychoanalysis, biosemiotics, and actor–network theory (ANT). The book is not intended to function as a means of entering the long debate on

DOI: 10.4324/9781003345572-1

the consciousness of machines or of answering the question of whether a machine can think. This does not mean that these debates are not present in the book; however, the book intends to provide a completely different point of view on these problems by proposing a new conceptual framework. Ultimately, it is a book about the relationship between humans and technology, and, more broadly, the meaning of technology.

This book first questions the usual way of understanding psychoanalysis. Through a new interpretation of Freud's works, I claim that Freudian psychoanalysis is a natural science based on a specific experimental method, such as biology or physics, the goal of which is to obtain knowledge of the human unconscious. This thesis has been improved and broadened by biosemiotics and ANT. I show that Freudian psychoanalysis is based on a biosemiotic approach to the mind and how ANT can integrate and develop this approach. The interaction of these three fields paves the way for a radical reformulation of the classic philosophical problem of the mind. This approach views the mind as a hybrid, a collective—a network of human and nonhuman actants in constant interaction with one another. The mind is a topology of networks in which human brains, bodies, and artifacts play equal roles. Unlike the extended mind approach, this book claims that artifacts are not passive extensions of human cognitive capabilities, but that they have "programs of action" and "counter-programs of action," and actively negotiate with humans and other nonhuman living beings. *If* the mind is distributed among things, and *if* things actively shape the mind, then consciousness and the unconscious are also distributed qualities and concern what Bruno Latour calls the "translation processes" in the actor network. There is no consciousness or unconscious but only certain assemblages of humans and nonhumans (i.e., certain networks and translation processes).

Psychoanalysis allows us to extend the ANT and to explore a new territory, that of *unconscious networks*, which refers to the set of invisible interactions of the actants of a particular network. This also allows us to extend the Freudian concept of resistance and repression to artifacts. Technology shapes the human unconscious, just as the human unconscious shapes technology. The ego, id, and superego are no longer just regions of an individual's abstract mind but parts of a collective, which itself comprises a set of human and nonhuman actants. The philosophical consequences of this thesis are remarkable.

On this theoretical basis, the book develops a new methodology, *technoanalysis*, and demonstrates its application to AI systems. Technoanalysis is a type of reverse engineering and interactive design; it involves the analysis of the machine as a network and identifies the internal resistances of this network—what I will call anti-mediations. The analysis of the resistances in the machine allows us to understand whether and how these resistances are connected to human drives and what effects they have. Technoanalysis promotes a novel understanding of some classic AI problems and makes possible a new awareness of the role of technology in the Anthropocene Age.

The Collingridge Dilemma and the Problem of the Control of AI

Can we control our technology? Can we get it to do what we want, and can we avoid its unwelcome consequences? The earlier assumption that technological evolution would automatically lead to significant social and human progress can no longer be sustained. The ambivalence of technology has become a standing topic in public, philosophical, and scientific debates. The scientific discussion about how to acquire and establish orientational knowledge for decision-makers facing the ambivalence of technology is divided into two branches: the ethics of technology and technology assessment (Grunwald 1999, 2018). These two branches are based on different assumptions concerning how to orient technology policy: the philosophical ethics branch, of course, emphasizes the normative implications of decisions related to technology and the importance of moral conflicts, while the technology assessment branch relies mainly on sociological or economic research.

The problem of evaluating and controlling technological development is at the heart of the so-called "Collingridge dilemma," which can be formulated as follows:

> attempting to control a technology is difficult, and not rarely impossible, because during its early stages, when it can be controlled, not enough can be known about its harmful social consequences to warrant controlling its development; but by the time these consequences are apparent, control has become costly and slow.
>
> (Collingridge 1980, 19)

For David Collingridge, technological development is always faster than the ability to understand its social effects. This asymmetry creates a strange effect: when changing a technology is simpler, especially at the beginning of its development, it is not perceived as a necessity, but when change is perceived as a necessity, it is no longer simple—it has become expensive and dangerous. "It is just impossible to foresee complex interactions between a technology and society over the time span required with sufficient certainty to justify controlling the technology now, when control may be very costly and disruptive" (Collingridge 1980, 12). This same asymmetry is evoked by Stiegler (1998): "Technique evolves faster than culture . . . everything happens as if time jumped out of itself . . . go faster than time" (18). We "do not immediately understand what is being played out in technics, nor what is being profoundly transformed therein, even though we unceasingly have to make decisions regarding technics, the consequences of which are felt to escape us more and more" (Stiegler 1998, 22).

The dilemma is not about technological development itself but about the perception that humans have of it and the awareness of its limits and effects. The technological development we produce exceeds our level of awareness and knowledge, and this affects our ability to forecast the social implications of technology: "A technology can be known to have unwanted social effects only when these effects are actually felt" (Collingridge 1980, 14). Why is it that, as technologies develop and become diffused, they become ever more resistant to controls that seek to alleviate their unwanted social consequences? To solve the dilemma, Collingridge (1980) develops a reversible, flexible decision-making theory that can be used when the decision-maker is still ignorant of the effects of a technology. His claim is that the essence of the control of technology is not in forecasting its social consequences "but in retaining the ability to change a technology, even when it is fully developed and diffused, so that any unwanted social consequences it may prove to have can be eliminated or ameliorated" (20–21). The important thing, for Collingridge, is to understand how to make the decisions that influence technological development in such a way as not to remain prisoners of them.

Faced with the technological development of the last 30 years and with pervasive phenomena, such as that of AI, Collingridge's theory of decision-making appears inadequate. Since Nick Bostrom (2014), the problem of controlling AI has become an important subject of discussion for many scholars. The prevailing tendency is to link the problem of control to the concept of *singularity*, which refers to the moment in human civilization when AI will overtake human intelligence. The problem of control is then interpreted as the problem of how to prevent superintelligence from harming human beings. For example, governments could use very powerful AI systems to carry out massacres, or a super system could slowly take control of the internet and global economy to manipulate the behaviors and opinions of billions of people (see the "Tale of the Omega Team" in Tegmark 2017, 3–21).

Bostrom (2014) depicts a dark future for humanity. He believes that the creation of a super-intelligent AI system could lead to the extinction of humankind. The risk involved in the creation of a superintelligence is that it would be operating at a speed and scale unfathomable to humans. This could initiate an intelligence explosion on a digital timescale of a millisecond so powerful that it could accidentally or deliberately destroy humanity. Bostrom not only contemplates the possibility of malicious applications of AI—such as hacking military devices; creating nano-factories, which can be distributed in undetectable concentrations, that manufacture killing devices on command; and duping humans into doing the "dirty work"—but also envisions a scenario in which AI achieves world domination and humans are useful only as raw materials. "Brains, if they contain information relevant to the AI's goals, could be disassembled and scanned, and the extracted data transferred to some more efficient and secure storage format" (Bostrom 2014, 118).

However, there is another interpretation of the control problem that has nothing to do with the singularity and destiny of humanity but, in general, has more to do with the relationship between humans and machines. This much less alarmist interpretation shows that the control problem is essentially a communication problem and is related to the very definition of AI. As Russell (2019) claims, "If we build machines to optimize objectives, the objectives we put into the machines have to match what we want, *but we do not know how to define human objectives completely and correctly*" (170, emphasis added). Human beings put their goals into the machine, but this is exactly the problem. Humans want the machine to do what we want, "*but we do not know how to define human objectives completely and correctly*," and we often act in ways that contradict our own preferences. Humanity is not a single, rational entity but "is composed of nasty, envy-driven, irrational, inconsistent, unstable, computationally limited, complex, evolving, heterogeneous entities. Loads and loads of them" (211).

The main challenge is to understand the nature of our goals and preferences. In Russell's (2019) view, "Preference change presents a challenge for theories of rationality at both the individual and societal level. . . . Machines cannot help but modify human preferences, because machines modify human experience" (241). How can we communicate our needs, values, and preferences to AI systems? This is a crucial problem in our world, where the influence of AI-based technologies is growing enormously. Unconscious dynamics influence AI and digital technology in general, and understanding them is essential to ensuring that we have better control of AI systems.

Wooldridge (2021, 240–241) criticizes the concept of singularity in Kurzweil's classical interpretation. His thesis is simple but effective: greater computing power does not imply greater intelligence. Increased computing power is a necessary but insufficient condition for the realization of an artificial general intelligence (AGI) or a superintelligence. This means that, even if we were to have very powerful computers, an AGI could still be a long way off. Starting from this premise, Wooldridge (2021) argues that the control problem is a communication problem intrinsic to AI. In other words, the essence of AI is based on a fundamental problem: Can we reduce what we call *intelligence* to software, and therefore to a program, or to a series of instructions that can be understood and executed by a Turing machine? There are different techniques and methods of computation and programming, so which is best for simulating human intelligence? In other words, if we want an AI system to act on our behalf, then we need to communicate to it what we want. However, "this turns out to be hard to do, and if we don't take care when we communicate our desires, then we may get what we asked for, but not what we actually wanted" (Wooldridge 2021, 239). How can we translate our desires and values into software and then into machine language—that is, into strings of ones and zeros?

There are two conditions to doing this: (1) we must understand what we really want from AI and (2) we must know how to communicate this to understand the connection and translatability between emotions and mathematics. However, we often do not know what we really want, we have contradictory desires, we do not really understand the consequences of our choices, or we are influenced by unconscious dynamics that we do not control. The lack of common values and norms makes normal communication with AI almost impossible; common sense is essential for understanding the meaning of our language, so AI may not understand the intentions hidden in our messages. A communication error could have dramatic consequences. For instance, the AI system may do what was intended but not in the way anticipated—a robot may be asked to ensure that a house is not burgled, so it burns it down.

An example of the extreme difficulty of this problem is the impossibility of translating Asimov's laws into an AI system (Wooldridge 2021, 245–246). We cannot translate ethical principles into a code. Therefore, the problem of control cannot be solved in ethical terms. This is not an ethical problem. I would say that, in general, we can distinguish two aspects of the control problem: (1) a psychological aspect that concerns the human—machine relationship and human projections onto the machine and (2) a linguistic aspect, whereby software is understood as a fundamentally symbolic mediation. An approach based on psychoanalysis could be of great help in seeking mediation between these two aspects. This is done by clarifying the desires and needs that move us, as well as the projections in which they are involved.

The problem of the control of AI can be seen as a branch, or an evolution, of the Collingridge dilemma. According to Kudina and Verbeek (2019), there are three types of approaches to the dilemma:

1 Anticipating developments (e.g., see Grunwald 2009); this approach is also called *constructive technology assessment* (see Rip, Misa, and Schot 1995).
2 Regulating developments; see the sociotechnical experimentation by van de Poel (2013). Many approaches to the ethics of technology can be connected to this approach. It is also Collingridge's (1980) approach.
3 Dynamism of the interaction, "which investigates how technologies mediate human practices, perceptions, and interpretations" (Kudina and Verbeek 2019, 294). This approach, inspired by post-phenomenology, is more dynamic, as it focuses on the interaction between humans and technology.

The present book intends to develop the third type of approach through the conceptual tools of psychoanalysis and ANT. My assumption is that solving the Collingridge dilemma requires the ability to answer two questions: What does technological development mean for human identity? What is the relationship between technology and the human mind?

I summarize the "plot" of the book in the following way:

Objective: This book aims to present a new conceptual framework for solving the control dilemma of Collingbridge. Understanding the social effects of technological development is essential for assessing the ethical impacts of technology.

Method: Concepts and methods drawn from psychoanalysis, biosemiotics, and ANT are applied to published sources, and case studies from the fields of social robotics and AI are analyzed.

Results: A new investigation method, *technoanalysis*, is proposed. This method offers a new multidisciplinary and systemic approach to the Collingridge dilemma based on the dynamism of the interaction between the human mind and technology. Technoanalysis allows us to understand how unconscious dynamics influence and shape technology and how, in turn, technology influences and shapes the unconscious. Understanding the relationship between the unconscious and technology is essential to controlling technological development. There is a vast body of literature on the relationship between psychoanalysis and AI; however, no one has ever thought of unifying psychoanalysis, biosemiotics, and ANT to solve the Collingbridge dilemma until now.

The Theoretical Framework of This Book

> There is today a conjunction between the question of technics and the question of time, one made evident by the speed of technical evolution, by the ruptures in temporalization (eventization) that this evolution provokes, and by the processes of deterritorialization accompanying it. It is a conjunction that calls for *a new consideration of technicity.*
>
> (Stiegler 1998, 17)

No thinker has reflected on the essence of technology more than Bernard Stiegler. To answer the question of what technology is Stiegler asserts that technology is inorganic matter organized in such a way as to constitute human temporality and spatiality. Stiegler claims that there is a profound continuity between biological and technical life; there is an internal technical tendency to biological life: "As a 'process of exteriorization,' technics is the pursuit of life by means other than life" (Stiegler 1998, 17). In *Technics and Time*, he writes:

> The human is a technical being that cannot be charactetized physiologically and specifically (in the zoological sense), for a diversity of human facts ruins the possibility of such scientific satisfaction related to the knowledge of the human qua human, and not only qua living being. . . . The evolution of the "prosthesis," not itself living, by which the human is nonetheless defined as a living being, constitutes the reality of the human's evolution, as if, with it, the history of life were to continue by means other than life: this is the paradox of a living being characterized in its forms of life by the nonliving—or by the traces that its life leaves in the nonliving.
>
> (Stiegler 1998, 50)

We must therefore think of a new relationship between human beings and technics beyond the rigid dualisms consecrated by tradition, such as ends/means or form/matter. Technology is mediation, no doubt, and yet, to say this is to say too little. Such a view focuses only on the function of the technology, its purpose—what it does, not what it is. Furthermore, reducing technology to mediation involves the risk of reducing it to a mere application of science or a practical end. What is the ontological dimension of technology? What is its relationship with human beings? Stiegler gives a convincing answer to this question, too: technology is memory (cognitive, bodily, behavioral, etc.) externalized, materialized, but not in the sense of a simple extension. The human being is constituted by an externalization of memory through technics.

> The zootechnological relation of the human to matter is a particular case of the relation of the living to its milieu, the former passing through organized inert matter—the technical object. The singularity of the relation lies in the fact that the inert, although organized, matter qua the technical object itself evolves in its organization: it is therefore no longer merely inert matter, but neither is it living matter. It is organized inorganic matter that transforms itself in time as living matter transforms itself in its interaction with the milieu. In addition, it becomes the interface through which the human qua living matter enters into relation with the milieu.
>
> (Stiegler 1998, 49)

For this reason, according to Stiegler, technics in the proper sense is mnemotechnics, inscriptions, or forms of writing—the trace, following Derrida (1967). Materialized memory is a full-fledged agent—an external organ capable of profoundly modifying the same vital organs from which it was born. Technology is not an application of scientific theories or practical purposes but an autonomous reality, independent of scientific progress and deeply connected to biological life. Following Stiegler, I do not claim that the concept of mediation should be excluded from a philosophical understanding of technology, but that it should be used critically. It is necessary to distinguish levels in the philosophical discourse on technology: the Stiglerian concept of epiphylogenetic memory is placed on a transcendental level, while that of mediation is placed on an empirical level. An ontology of technology must hold these two levels together.

Stiegler's thesis is fascinating because it touches on a very profound point: technics has to do with life itself—that is, with the ability to remember, hold, and organize time—and therefore with what is more unique in human beings and even in all living beings. Technics arises from the need to strengthen and control individuation in the sense of Simondon (2005)—that is, the constitution of physical, living, psychic, and collective individuality. Stiegler, who is also a great critic of Simondon, connects technics, memory, and individuation; this is the core of his thought. These three elements define each other mutually.

An objection to this understanding could be that, by combining technogenesis and anthropogenesis, Stiegler lacks anthropomorphism. This is not a sustainable objection, however, as I believe that it is necessary to distinguish anthropomorphism from anthropocentrism. Technics are produced by humans; we do not know, at least so far, of technics that are not produced by human beings. As such, talking about technics means talking about the human being—or about some animal species; it is very different to have an anthropocentric vision of technology, that is, a vision that reduces technology to the will of the human being. In my opinion, it is plausible to maintain an anthropomorphic vision without being anthropocentric.

Starting from Heidegger's critique, Derrida's logic of supplement, and Leroi-Gourhan's research on paleoanthropology, Stiegler affirms that the origin of hominization involves the process of the externalization and reinternalization of memory. Technology is therefore strictly connected to time and memory; it is the inanimate, inorganic matter that produces human temporality (Stiegler 1998). The anthropological assumption of this thesis is that the human being is a defective being who needs prostheses to live, reproduce, socialize, express themself, and think; therefore, a movement of materialization of memory is needed to make up for the lack of human memory. Technology is not the oblivion of the difference between being and time, as Heidegger argued. Instead, it is the original condition of the possibility of time and of the relationship between time and being. For Stiegler, anthropogenesis and technogenesis are the same phenomena considered from two different points of view. The appearance of humans is the appearance of technology.

In more technical terms, hominization is *grammatization*—an expression that Stiegler takes up from Derrida (1967) and Auroux (1993). *Grammatization* is the process of the description, formalization, and discretization of human behaviors that allow their reproducibility by giving them materiality, spatiality, and the possibility of sedimentation.

> Grammatisation is the history of the exteriorisation of memory in all its forms: nervous and cerebral memory, first linguistic then auditive and visual, bodily and muscular memory, biogenetic memory. Thus exteriorized, memory becomes the object of socio-political and biopolitical controls through the economic investments of social organisations which thus retool psychical organisations by means of mnemotechnical organs, including machine tools (Adam Smith analyzed as early as 1776 the effects of the machine on the mind of the worker) and all the automats, including household equipment. This is why a thinking of grammatisation calls for a general organology, that is, a theory of the articulation of bodily organs, artificial organs and social organisations.
>
> (Stiegler 2006, 16)

I will later explore the link between grammatization and organology. For the moment, however, I want to underline the connection between the concept of grammatization and Stiegler's reinterpretation of Husserl's (1991) theory of retention. Through this reinterpretation, Stiegler elaborates the two key concepts of his work: tertiary retention and epiphylogenetic memory. Following Husserl, Stiegler distinguishes primary retention—the neurophysiological structures of an individual's perception and memory—from secondary retention—the psychic memory of an individual processed through the imagination. *Primary retention* concerns the present; it is the ability to hold onto a particular sound while listening to a melody and compare it with other sounds—the previous note remains present in the perception. *Secondary retention*, in contrast, is what we properly call the *past*, and it involves the ability to bring back a melody previously heard.

Further still, *tertiary retention* is the externalization of memory. Take, for example, a melody recorded by a phonograph. This form of remembering is neither a primary retention nor a secondary retention. It is something completely different. It is an exteriorization, a materialization, a supplement that allows us to re-listen to the same melody in an authentic way at different times. But there is more: according to Stiegler, tertiary retention is the condition of primary and secondary retention in the sense that, although it arises from them, it is nevertheless able to condition and modify them. The mnemotechnical supplements—in different degrees, corresponding to their historical evolution—condition and modify not only the selection and construction criteria of the memory in individuals (i.e., secondary retention) but also the neurophysiological structure of the human perception/memory system, that is, the so-called *phylogenetic memory* (i.e., primary retention). This is a recursive evolutionary conditioning. Therefore, Stiegler differs profoundly from Derrida (1967, 1974); the *différance* is not the general structure of retention—that is, the condition of the possibility of memory—but the effect of the supplement (i.e., writing) on the neurophysiological structure of the human perception/memory system.

From this point of view, Stiegler is closely aligned with the Material Engagement Theory (MET) of Malafouris (2013) and the ANT of Latour; both theories give nonhuman agents, or "material culture," an active role in the constitution of psychosocial reality—the mind is shaped by the objects it uses and with which it relates. Similarly, for Stiegler, nonhuman agents (e.g., the phonograph, material culture, or retention devices) actively modify human agents (and living agents in general) in the course of a recursive evolution—an evolution that constantly returns to itself and changes itself. Memory is the very condition of life, but this condition is technically defined. The biological architecture of the living, even the genetic program that guarantees the identity and conservation of the species in the course of evolution, is technically mediated, influenced, and defined.

This is the profound meaning of the concept of epiphylogenetic memory, which is the direct consequence of the concept of tertiary retention.

> Epipylogenesis, a recapitulating, dynamic, and morphogenetic (phylogenetic) accumulation of individual experience (epi), designates the appearance of a new relation between the organism and its environment, which is also a new state of matter. If the individual is organic organized matter, then its relation to its environment (to matter in general, organic or inorganic), when it is a question of a who, is mediated by the organized but inorganic matter of the organon, the tool with its instructive role (its role qua instrument), the what. It is in this sense that the what invents the who just as much as it is invented by it.
>
> (Stiegler 1998, 177)

Epiphylogenesis is a recursive and evolutionary phenomenon—a union between phylogeny and epigenesis. This phenomenon produces a new state of matter: organized inorganic matter. For Stiegler, this is technology. Vital organs are the result of an evolutionary process through which certain matter receives a certain level of organization. By evolving, the vital organs produce new organs, external organs—that is, non-living matter endowed with a certain degree of organization. In turn, these external organs condition and transform the internal organs (i.e., the organs inside the body). It is a recursive and recapitulative movement.

There are two important consequences of this thesis. The first is that biological evolution is not a linear process but a recursive one, in which the results influence the premises by redefining them. In other words, the biological evolution of humans is not simply a natural fact but can be conditioned by cultural and technical factors; therefore, it is possible to intervene in the human neurophysiological structure to modify it. The second consequence of Stiegler's thesis is that it eliminates the classic distinction between inside and outside. The interiority of the individual is defined by the exteriority of the technology. Epiphylogenesis is a movement of exteriorization that folds in on itself, generating new forms. This movement is at the root of hominization and defines the very evolution of humans. The human being is defined by this tendency to externalize/reinternalize their body, drives, consciousness and unconscious, cognitive faculties, and so on. However, nothing prevents us from extending this idea to other living forms with an adequate neurophysiological structure.

There is also a third consequence of the concept of epiphilogenesis which is fundamental to this book. Artifacts are materialized memory. For this reason, they speak of us and our history. They are layers of memories; some of them are conscious and others unconscious. One of the fundamental tasks of technoanalysis is to explore the most unconscious layers of these memories.

These are the premises of the philosophical program that Stiegler calls *general organology*, which is not only a descriptive theory but also a symptomatology. Biological life produces not only vital organs, internal to the human body, but also external, artificial, and social organs; all are the expression of a phase of precarious equilibrium of its evolution. *General organology* is therefore a method of analyzing the history and evolution of physiological organs, artificial organs, and social organizations based on the concept of epiphylogenesis. This means that organology establishes a relationship between the three types of organs such that a variation in one of them always involves variation in the others:

> "General organology" is a method of joint analysis of the history and future of physiological organs, artificial organs and social organizations. It describes a transductive relationship between three types of "organs": physiological, technical and social. The relation is transductive insofar as the variation of a term of one type always involves the variation of the terms of the two other types. A physiological organ—including the brain—does not evolve independently of technical and social organs. The psychic apparatus is not reducible to the brain and presupposes technical organs, artefacts that support symbolization and of which language is a case.
>
> (Stiegler 2009, 41)

Organology becomes pharmacology when it raises the question of the ambivalence and possible toxicity of grammatization. In this case, Stiegler reinterprets and extends Derrida's (1972) concept of pharmakon. Identifying the inherent risks of the grammarization of experience means establishing a therapeutic approach—that is, conceiving and developing a new industrial and educational model, as opposed to post-Fordist capitalism. Pharmacology is a positive enterprise: the concrete invention of unprecedented technological practices.

Grammatization and epiphylogenesis are the basis of psychic and collective individuation, a concept that Stiegler takes up from Simondon (2005). Individuation, or the constitution of individuality (human and otherwise), is a *becoming* without an end; every individuality, according to Simondon's thesis, is a precarious equilibrium between forces starting from a pre-individual state. Simondon clearly criticizes the hylomorphism and substantialism of the metaphysical tradition and replaces it with a model based on the tension between forces; the individual is what maintains itself in the tension. "The emergence of an individual within the preindividual being should be conceived in terms of the resolution of a tension between potentials belonging to previously separated orders of magnitude" (Combes 2012, 4). The individual is therefore always a metastable equilibrium, neither stable nor unstable. Stiegler criticizes Simondon by stating that the pre-individual condition and the transindividual processes that connect psychic and collective individuation are essentially technical—that is, forms of grammar and sets of tertiary retentions. Thus, technology is at the heart of identification. Psychic identification is originally collective and, precisely for this reason, technical; the secondary retentions must be stabilized and shared through grammatization. By doing this, Stiegler solves what he believes is one of Simondon's crucial problems: the lack of a discussion of the role of technical individuation in the link between psychic individuation and collective individuation.

Organology, in the sense understood by Stiegler, represents the conceptual and methodological framework within which the investigation of the present book is positioned. The problem of controlling technology and AI will therefore be reinterpreted in organological and pharmacological terms. *Technoanalysis* will be understood as the exploration of the technical unconscious that presides over the forms of psychic and collective individuation occurring today. Nevertheless, as shown in Figure 0.1, I also intend to develop the conceptual framework provided by Stiegler through the integration of ANT and MET. Furthermore, in my opinion, the biosemiotics developed by Sebeok, starting from Peirce—as well as that of Barbieri, despite the differences—represent important resources to explain the emergence of technology, as they allow us to overcome the traditional dualism of matter/meaning.

From the point of view of organology, AI is the most extreme form of epiphylogenesis. For Stiegler (2018, 1),

> what we today refer to as artificial intelligence is a continuation of the process of the exosomatization of noesis itself, such as it begins firstly with fabricating exosomatization, making things by hand, and continues with hypomnesic exosomatization, as that which makes it possible to access lived experiences of memory and imagination.

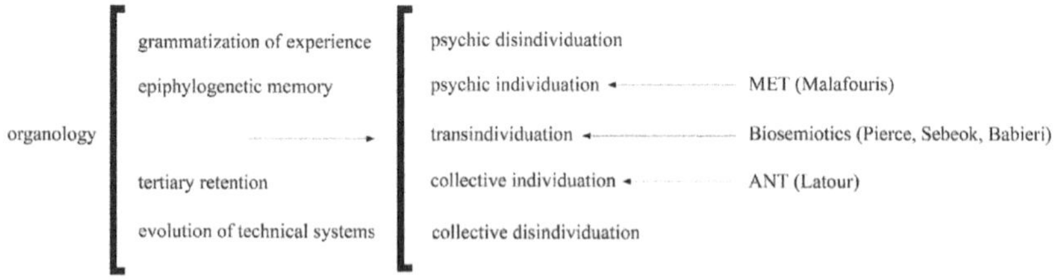

Figure 0.1 The theoretical structure of this book: Stiegler's organology, improved by MET, ANT, and biosemiotics. Technoanalysis is an extension of organology.

AI is nothing more than the externalization process of intelligence that Stiegler calls *noesis*—that is, psychic intelligence that is endowed with a certain kind of conscious and unconscious and that, therefore, cannot be reduced to cognitive functions alone. This is not the same as arguing that machines have a consciousness and an unconscious. Instead, it is equivalent to saying that AI is a new form of tertiary retention—that is, a grammatization of the human psychic identity—and that it therefore has the ability, like all technical objects, to influence and shape this identity. Our unconscious influences the AI, which, in turn, can influence and shape the unconscious itself. This means that AI can also be artificial stupidity: "Artificial stupidity, then, is what persists in accelerating entropy instead of deferring it and does so by destroying knowledge" (Stiegler 2018, 7).

What Is AI?

Trying to define the boundaries of an ever-expanding field like AI would mean writing another book. This book was born from the awareness that, today, we live in the AI ecosystem and that defining the boundaries of this ecosystem from the inside is increasingly complex. AI is a general term that brings together very different things. It is a set of technologies whose main feature is that their behavior can be associated with the idea we commonly have of human intelligence. According to Wooldridge (2021),

> The long-term dream of AI is to build machines that have the full range of capabilities for intelligent action that people have—to build machines that are self-aware, conscious and autonomous in the same way that people like you and me are" (2).

This type of AI, which is generally called AGI, is a dream that, for the moment, is destined to remain in the world of fantasy. To put it simply, AI systems like HAL or the Terminator do not exist; we cannot yet build

AI that is able to reproduce all the aspects of human intelligence (Floridi 2021). Instead, there is a set of technologies that can reproduce some human cognitive abilities, such as learning, speaking, and recognizing images. As Wooldridge (2021) notes, "The mainstream AI researcher today is focused around getting machines to do specific tasks which currently require human brains (and also, potentially, human bodies), and for which conventional computing techniques provide no solutions" (3). In this book, when I talk about AI, I am referring to this second type of AI, the so-called narrow or weak AI. I limit myself to considering AI as a set of technologies whose goal is to simulate some human faculties and collaborate in an active and participatory way with humans. This, however, does not mean that fantasies and projects involving AGI cannot teach us about AI and the way in which the collective imagination perceives and thinks about AI. AGI projects and the imagery connected to them can influence the development of weak AI; there is a dialectical relationship between the two sides of the problem of AI. AI is therefore an extremely complex phenomenon. This complexity makes it a multifaceted object of study that requires a highly interdisciplinary methodology.

According to a leading textbook on AI by Russell and Norvig (2016), intelligent systems can be classified into the following four systems:

1. Systems that think like humans, where the focus is on cognitive modeling (e.g., cognitive architectures and neural networks);
2. Systems that act like humans, with a focus on simulating human activity, and which are evaluated by applying Turing-like tests (see Turing 1950);
3. Systems that think rationally by using logic-based approaches to model uncertainty and deal with complexity (e.g., problem solvers, inference, theorem provers, and optimization);
4. Systems that act rationally, where the focus is on agents that maximize the expected value of their performance in their environment.

AI can also be classified according to the methods used, such as symbolic AI (using logic), connectionist approaches (inspired by the human brain), evolutionary methods (inspired by Darwinian evolution), probabilistic inference (based on Bayesian networks), and analogical methods (based on extrapolation). As Dignum (2019) has stated, "Building intelligent machines has many different facets, including understanding language, solving problems, planning, recognizing images and patterns, communicating, learning, and many more" (12). Different areas of research are characterized by the means they employ to achieve these aims, and "this is why sub-fields of AI, such as machine learning, natural language understanding, pattern recognition, evolutionary and genetic computing, expert systems or speech processing, sometimes have very little in common" (Ibid.).

Despite this multiplicity of techniques and applications, we can generally distinguish between two major approaches to the creation and development of AI.

The top-down approach, also known as symbolic AI, or Good Old-Fashioned AI, attempts to explicitly represent human knowledge in a declarative form (i.e., as facts and rules). This approach focuses on the translation of often implicit or procedural knowledge into formal knowledge representation rules to make deductions, derive new knowledge, and inform action. Top-down approaches are grounded in the notion that intelligence can be reproduced by rational logic. However, several mundane tasks (for instance, to make a cake or to move objects) are not amenable to be formalized by logical techniques.

The bottom-up approach, based on learning from experience, models intelligence without explicit representations of knowledge. Sometimes referred to as sub-symbolic or connectionist approaches, these systems loosely take inspiration from how the brain works and are generally associated with the metaphor of a neuron. Indeed, such systems serve as the basis for neural network architectures. These approaches require large amounts of data and are particularly suitable for solving specific problems, such as recognizing natural language or images. Their success depends on the availability of data and computational power. These approaches are mainly those of machine learning and deep neural networks. The goal of these systems "is to have programs that can compute a desired output from a given input, without being given an explicit recipe for how to do this" (Wooldridge 2021, 169). Despite the successes of these techniques, there are also critics (Langley 2011; Pearl and Mackenzie 2018; on machine learning, see also Carbonell et al. 1983).

A purely technical picture of AI, however, would be too partial. As Crawford (2021) points out, AI is not only a bundle of software. What we do not see "beyond" the software is an immense mining industry that exploits energy and mineral resources, cheap labor, and, finally, data that are chosen by users according to their wishes. AI is not just what we see on a screen. Instead, it is an immense supply and logistics chain that envelops the entire planet with significant social, political, and environmental consequences. AI "is both embodied and material, made from natural resources, fuel, human labor, infrastructures, logistics, histories, and classifications"; it is not "autonomous, rational, or able to discern anything without extensive, computationally intensive training with large datasets or predefined rules and rewards" (Crawford 2021, 8).

The spread of AI systems has profound social, psychological, and cultural consequences throughout the world. The ubiquitous spread of software algorithms, deep learning, advanced robotics, accelerating automation, and machine decision-making, when contextualized in terms of the global digital distribution and use of internet-connected devices that generate massive quantities of data, generates complex new systems and processes with multiple impacts across social, cultural, political, and institutional life. "Lifestyles permeated by AI are intricately interwoven with extensive and highly intensive complex digital systems" (Elliott 2018, xviii).

This inevitably raises ethical problems. Having delegated most of the decisions in sensitive fields, such as healthcare, the job market, finance, and courts, to AI systems implies the need to provide these systems with ethical analysis capabilities. How many decisions do we want to delegate to AI? Who is responsible when something goes wrong? A discussion of AI ethics cannot be like other traditional types of ethics because it must face new kinds of problems; this is because "the common approaches may not be sufficient, primarily due to the transformational nature of AI within science, engineering, and human culture" (Powers and Ganascia 2020, 28; see also Coeckelberg 2020a, b). From another point of view, an ethical approach to AI must inevitably be connected to design and therefore develop on three levels: ethics in design, ethics by design, and ethics for design (Dignum 2019).

Finally, what we call "AI" in this book cannot be considered a single phenomenon, but a family of phenomena that are (a) historical, in the sense that they express or depend on a specific phase of post-Fordist capitalism; (b) technological; and (c) geographically located (e.g., infrastructure networks, supply and logistics chains, mines, and data centers); they also include (d) imaginaries of the present and above all of the future of humanity, as demonstrated by the concept of singularity (Kurzweil 2005; Tegmark 2017) or the explosion of intelligence (Bostrom 2014). AI is an integral part of how the twenty-first-century human being describes itself and its future.

Why Psychoanalysis?

Why do I choose psychoanalysis to study AI and no other psychological doctrines or forms of psychotherapy? What kind of psychoanalysis do I use? These are important questions because they allow me to clarify some central methodological aspects of my research. I consider the Freudian psychoanalytic method the only real science of the mind because it solves the crucial problem of suggestion—what psychotherapy cannot do, as I will explain. This does not mean, however, that I completely reject any other approach to psychoanalysis that is not Freudian. I now want to clarify and develop four points:

1 The definition of psychoanalysis assumed in this book;
2 The relationship between psychoanalysis and medicine;
3 The relationship between psychoanalysis and psychotherapy;
4 The ethics of psychoanalysis.

The first aim of this book is to show how the widespread idea of psychoanalysis is wrong and that Freud has been misinterpreted for a long time. Psychoanalysis is not an object of cultural curiosity that belongs to the past. Instead, it is a rigorous scientific discipline. Providing a definition of psychoanalysis is a very complex task, given the nature of the object; psychoanalysis is, in fact, a difficult, long work and requires great skill and dedication. Undoubtedly, the current state of psychoanalysis in the world does not help. Today, the title of psychoanalysis is claimed by many different schools or organizations, which refer to approaches and theories that often have little, if anything, in common. This list includes Kleinians, Bionians, Winnicottians, Jungians, Adlerians, psychoanalysis of the self (Kohut), feminist psychoanalysis (Irigay and Benjamin), interpersonal psychoanalysis (Sullivan and Mitchell), and neuropsychoanalysis (Solms). There are also Lacan and a thousand of his mimics who have developed his ideas in philosophy and beyond. Freud's ideas have also known thousands of evolutions, deformations, and transformations in the human sciences and psychotherapy. It would be too lengthy and futile a task to quote the literature on the subject (see Ellenberger 1970; Mitchell and Black 1995; Roazen 1990). Today, there is no psychoanalysis, but psychoanalyses. Psychoanalysis is a patchwork without unity, without a shared method, and its theorists contradict each other; thus, the field has a rather declining reputation. As Baldini notes, "All the new hypotheses proposed were advanced on a narcissistic basis, not on an objective basis" (Baldini 2021, 25; my translation).

This book takes a clear position. I define psychoanalysis as a science based on a specific experimental method. Its purpose is to secure objective knowledge of unconscious processes. I choose to define psychoanalysis this way because this is Freud's position, which has unfortunately been forgotten by many psychoanalysts.

Like physics or biology, psychoanalysis has a clear, standard definition. In 1922, in the preface to a text on sexology by Max Marcuse, Freud wrote:

> Psychoanalysis is the name (1) of a procedure for the investigation of mental processes that are almost inaccessible in any other way, (2) of a method (based upon that investigation) for the treatment of neurotic disorders and (3) of a collection of psychological information obtained along those lines, which is gradually being accumulated into a new scientific discipline.
>
> (Freud 1955, 235)

Freud states that psychoanalysis is first a method of investigation and second a treatment of neurotic disorders (i.e., a therapeutic practice). Psychoanalysis, therefore, has a cognitive, and not a clinical, purpose. Therapy depends on the method of investigation, not the other way around. This means that the therapy is entirely conditioned by the method of investigation and must depend on the implementation of the latter in the different situations in which a subject is treated—that is, in certain situations, the method of investigation can also become a treatment that cures psychoneuroses. However, the fact that it has therapeutic effects does not mean that these effects constitute the main purpose of the investigation. This is an essential point: psychoanalysis is not a therapy; its main purpose is not to cure the patient, and therefore the fact that the method of investigation becomes therapy is completely secondary. This is why we cannot reduce psychoanalysis to psychotherapy. Freud himself remarks that "psychoanalysis has never set itself up as a panacea and has never claimed to perform miracles" (Freud 1955, 250). He then adds the following:

> In one of the most difficult spheres of medical activity it is the only possible method of treatment for certain illnesses and for others it is the method which yields the best or the most permanent results—though never without a corresponding expenditure of time and trouble. A physician who is not wholly absorbed in the work of giving help will find his labors amply repaid by obtaining an unhoped-for insight into the complications of mental life and the interrelations between the mental and the physical. Where at present it cannot offer help but only theoretical understanding, it may perhaps be preparing the way for some later, more direct means of influencing neurotic disorders.
>
> (Freud 1955, 250)

Freud understands psychoanalysis as an empirical science, in the same vein as physics or chemistry, based on a specific method of verification, which possibly, under certain conditions, can allow for the treatment of neurotic disorders. The object of this science is the objective knowledge of the unconscious. Any theory that abandons this starting point, or that betrays it, cannot be called psychoanalysis. One of the purposes of this book is to develop and strengthen this thesis.

Much of the history of post-Freudian psychoanalysis has been characterized by the desire to reduce psychoanalysis to psychotherapy or a psychiatric superspecialization, especially in the United States (Dalto 2021, 180). This phenomenon has caused a progressive erosion of Freudian theoretical heritage and of the very identity of the discipline founded by Freud. A clear demonstration of this is found in Mills (2000), who presents a very broad conception of psychoanalysis applicable to very different settings. For Mill, what truly characterizes psychoanalysis is the analysis of transference. The concept of suggestion is present, but it is not problematized, and there is no method of verification. Mills does not distinguish between interpretation and construction; he conflates these aspects and reduces them to the analysis of transference.

Considering psychoanalysis as a medical treatment is wrong for two reasons:

1 While in psychoanalysis, the cognitive moment is independent of therapy, in medicine, it is preliminary to therapy because therapy is the goal that conditions everything else.
2 The object of psychoanalysis is completely different from that of medicine. The subject, what psychoanalysis is about, is a special object—it cannot be treated like the heart or the lungs. Therefore, psychoanalysis and medicine require very different methods of verification.

These differences become more evident when psychoanalysis and psychotherapy are compared. As I stated earlier, psychoanalysis is not a form of psychotherapy, nor can it be reduced to psychotherapy—to do so would mean betraying the identity of psychoanalysis. From the point of view of psychoanalysis, psychotherapy is vitiated by heavy methodological and theoretical flaws that prevent it from being defined as a science (see Ceschi 2021; Evers 2018; Lambert and Bergin 1992). Here, I mention three of these flaws:

1 *The problem of equivalence*: All psychotherapies (behavioral, cognitive-behavioral, Gestalt, group, psychosomatic, psychodynamic, constructivist, imaginative, etc.) are equivalent, in the sense that it is not possible to identify statistically significant differences in efficacy. The techniques and theories used cannot be considered responsible for the results, whether positive or negative (Benedetti 2015, 246). Therapy outcomes vary more due to variables related to the people involved than to methodological and theoretical approaches. In the psychotherapy context, the so-called "alliance" with the therapist—I mean the relation with the doctor—seems to be the real discriminating factor. To put it simply, it is not possible to prove that one approach is better than another. To this are added other technical problems, such as the weight of the therapist's prejudices, the difficulty of having unique measurement scales, and the lack of shared definitions.
2 *The placebo problem*: How can the therapist be sure that the improvement achieved by the patient during therapy is not just the effect of a suggestion and therefore nothing more than a temporary and unstable change? In psychotherapy, there is a tendency to assume the double-blind method of verifying theoretical hypotheses, which is very close to the approaches used in pharmacology and medicine. Medicine eliminates the placebo problem by administering the same active ingredient to a whole group of people and the same placebo to another group without either knowing which substance they took or

without even the doctor knowing much about it. This extra-clinical procedure (because it separates experiment and clinic) leads to statistical evidence: we know how likely a drug is to act effectively. Pharmacology works in this way, and psychotherapy follows this model without posing any problems (see Rosenthal and Frank 1956; Wampol et al. 2005; Kirsh et al. 2016; Blease and Kirsh 2016), except for one: it is not possible to apply the double-blind control method in psychotherapy because, in this case, it is impossible to administer the same active ingredient to each member of a group. This is due to the specific nature of the object of psychotherapy, which is the human subject. Therefore, a psychotherapist cannot be sure that the improvement achieved by the patient during therapy is not just the effect of a suggestion and therefore nothing more than a temporary change. According to Ceschi (2021), "In psychotherapy every psychotherapeutic treatment consists of a long series of interactions between patient and therapist and the suggestive aspects can therefore enter at any moment both real therapy and phantom control therapy" (50; my translation). Gaab et al. (2016) arrived at the same results. The problem of placebo/suggestion is a fundamental aspect. Because it follows a sanitary and medical-pharmacological model, psychotherapy has no method of verifying whether the improvements in the patient, which are the goal of the cure, are only due to a suggestive effect and therefore by nature inconsistent and temporary, or to the methodological approach and to the underlying theory. The conclusion is that psychotherapeutic intervention is indistinguishable from mere suggestions. As we will see in Chapter 1, Freud created and developed an intraclinical method (because it does not separate experiment and clinic) for the experimental verification of the patient's improvements that solve the placebo problem. This intraclinical method is able to distinguish those improvements due to the suggestion from those due to the analyst's intervention and its theory. We could almost say that psychoanalysis consists of the discovery of this method. This distinguishes psychoanalysis and psychotherapy; for the former, the improvement of the patient is a problem, or a starting point to be explained, and not the goal of the investigation.

3 *The problem of the reproducibility of the results*: This is another huge problem that the scientific community has been wondering about for a long time (see Open Science Collaboration 2015; Baker 2016). Experimental science implies the possibility of reproducing the results of experiments. However, this is impossible in psychotherapies, where the experiment and its evaluation cannot ignore subjective variables. The psychotherapeutic situation is *sui generis*: an object (i.e., the human subject) is investigated by a researcher who is himself–herself a subject. Most psychotherapy experiments cannot

> be reproduced or objectively evaluated. The Freudian method also solves this problem.

Psychoanalysis is incompatible with any type of psychotherapy. The purpose of psychoanalysis is to secure knowledge (i.e., the broadening of the patient-analysand's knowledge of herself–himself and of the genesis and development of his–her pathology). In other words, the analysis must establish the best conditions for the functioning of the ego. The primary purpose of psychoanalysis is not to alleviate the suffering of the patient-analysand (i.e., to heal him–her by eliminating the symptoms of the disease). The analyst offers a *metatherapy* (Baldini 2021, 15) in the sense that she–he builds the conditions of the possibility of a self-therapy that must be entirely in the hands of the patient-analysand. From this point of view, psychoanalysis unmasks the inconsistency of psychotherapy as a science. There are dozens of different types of psychotherapeutic and psychological approaches in the world today, the foundations of which are not scientific but only social or academic:

> When we study Western psychotherapies, we generally find an "entity" that describes itself as objective and scientific, or methodologically aware and epistemologically founded, but we must ascertain whether this vision is not simply a compliant self-narration (as in astrology) or if it actually corresponds to something substantial (as in biomedicine).
>
> (Salvador 2021, 159–160; my translation)

This judgment is shared by Luborsky et al. (1997) and Nathan and Zajde (2013), who have developed a criticism that is not simply generalist, like that of Szasz or Foucault, but methodological and conceptual.

The classic objection made to these theses is that the patient does not have a great interest in the knowledge of the unconscious. The patient wants to heal. What she–he really wants is to be freed from suffering, and for this reason, psychoanalysis must become a sort of psychotherapy. Many psychoanalysts hold this position (for instance, Fink 1999 and Cornell 2019).

This objection is patently nonsensical because it forgets a key point: the patient wishes not to heal at least as much as she or he wishes to heal. This is an aspect that Freud highlights when speaking of the unconscious need for punishment and the compulsion to repeat. The patient does not want to heal because of his–her unconscious resistance to healing. It is not that there is no desire for healing; it is there, but it is not enough.

By renouncing therapy, psychoanalysis does not renounce efficacy. The analyst interprets the material provided by the patient and elaborates on what Freud calls "constructions," or theoretical hypotheses concerning the history of the patient and the origin of psychoneurosis.

The construction is then tested experimentally in the session with the patient to establish its objectivity. Having carried out such work, the analyst communicates the construction to the patient. Is the communication of the construction sufficient to produce changes in the patient, that is, to eliminate resistance? Resistances do not fall in front of the truth; for Freud, truth is always a correspondence of a theoretical hypothesis to a reality that has been verified. Thinking that truth is enough to eliminate resistance would mean falling back into an intellectualistic prejudice (Dalto 2021, 74–75). There are many resistances, and repression is only one type. Even if the repression falls away, the resistances of the Id (compulsion to repeat) and those of the Superego (unconscious sense of guilt) remain active, blocking the patient and preventing his–her real change.

By revealing the objective truth of the unconscious, the analyst has done his–her job—the analysis is over. Nevertheless, the analyst can still help the patient accept that truth and then choose to heal. As Baldini (2021) notes, "The therapeutic success of the analysis cannot be achieved by the analyst; however, it can be prepared" (21; my translation). The analyst implements what we can call pragmatics, or an education, not a therapy.

Consider the following passage from *On Beginning the Treatment*:

> The primary motive force in the therapy is the patient's suffering and the wish to be cured that arises from it. The strength of this motive force is subtracted from by various factors—which are not discovered till the analysis is in progress—above all, by what we have called the "secondary gain from illness;" but it must be maintained till the end of the treatment. Every improvement effects a diminution of it. By itself, however, this motive force is not sufficient to get rid of the illness. Two things are lacking in it for this: it does not know what paths to follow to reach this end; and it does not possess the necessary quota of energy with which to oppose the resistances. The analytic treatment helps to remedy both these deficiencies. It supplies the amounts of energy that are needed for overcoming the resistances by making mobile the energies which lie ready for the transference; and, by giving the patient information at the right time, it shows him the paths along which he should direct those energies. Often enough the transference is able to remove the symptoms of the disease by itself, but only for a while—only for as long as it itself lasts. In this case the treatment is a treatment by suggestion, and not a psychoanalysis at all. It only deserves the latter name if the intensity of the transference has been utilized for the overcoming of resistances. Only then has being ill become impossible, even when the transference has once more been dissolved, which is its destined end.
>
> (Freud 1958, 143)

In the first part of the passage, Freud clearly states that the desire for healing is insufficient. To carry out an analysis, something else is needed: a desire for knowledge. Only knowledge can make the difference and create the conditions for the patient to obtain the energy to cope with his-her illness and overcome it: "by giving the patient information at the right time, it shows him the paths along which he should direct those energies."

In the second part of the passage, Freud confirms what we have said so far: psychoanalysis is not a suggestion. Then, he adds the following:

> In the course of the treatment, yet another helpful factor is aroused. This is the patient's intellectual interest and understanding. But this alone hardly comes into consideration in comparison with the other forces that are engaged in the struggle; for it is always in danger of losing its value as a result of the clouding of judgment that arises from the resistances.
>
> (Freud 1958, 143)

The truth proposed by the analyst is a hypothesis on the historical truth of the psychic development of the patient. This historical truth is ineffective; alone, it cannot change anything. As Baldini (2021) and Dalto (2021) underline, the psychoanalyst must also provide a method to the patient—that is, a pragmatics of truth, or a "mental map of conduct" (Baldini 2021, 23; my translation):

> Thus the new sources of strength for which the patient is indebted to his analyst are reducible to transference and instruction (through the communications made to him). The patient, however, only makes use of the instruction in so far as he is induced to do so by the transference; and it is for this reason that our first communication should be withheld until a strong transference has been established. And this, we may add, holds good of every subsequent communication. In each case we must wait until the disturbance of the transference by the successive emergence of transference-resistances has been removed.
>
> (Freud 1958, 143–144)

Here, Freud means that during treatment, the desire for healing progressively diminishes, while the intellectual desire to understand the origin of one's illness grows. Intellectual intention leads to the possibility of healing, but it does not heal in itself. The psychoanalyst must fully respect the freedom of the patient; he–she must give him–her the theoretical tools and the method to heal, not effectively heal him–her. A pragmatics of truth, also called "post-education," must therefore be linked to the experimental verification of constructions.

This is connected to the ethical side of psychoanalysis. Psychoanalysis has, in fact, an ethical dimension, which entails the teaching of a subject's finitude. The neurotic is basically one who does not accept the finitude imposed by the surrounding world and refuses to limit the extent of his–her enjoyment. The analyst, in contrast, leads the neurotic back to the finitude of enjoyment; the analysis makes the patient able to accept his–her life and frustration. This is the improvement of the ego proposed by the analyst—the highest degree of autonomy in the Kantian sense. Aware of his–her finitude, the patient has full freedom to dispose of the libido as she–he wishes (Baldini 2021, 30–31). Resistances first cause cognitive impediments, such as the inability to remember or access the information necessary to know about the world and enjoy it. Fixations restrict our view of the world and our ability to live in it. The analyst operates on these fixations, brings them to light, and reveals their origin. In this process, he–she also assumes a pragmatic attitude (in a Peircean sense), insofar as she–he can give a method to the patient to freely choose to heal or not.

Literature Survey

The methodological premises I have just outlined differentiate this book from any previous attempt to use psychoanalysis to understand digital technology and AI. I do not attempt to draw a conclusion about the state of the art or to provide a kind of annotated bibliography in an effort to be as complete as possible. I think such tasks would be futile. It is much more fruitful to analyze a few important examples of books that have investigated the relationship between psychoanalysis and AI and to show the methodological diversity of this book.

The first scholar to recognize the importance of psychoanalysis for the study of AI and digital technology was the MIT philosopher and sociologist Sherry Turkle, author of vital and influential texts, such as *The Second Self* (1984), *Life on the Screen* (1995), and *Alone Together* (2011a). In these texts, Turkle analyzes the transformations of the personalities of children and adults in contact with new digital technologies. Turkle is mainly interested in the "psychology of people's relationships with computers" (1984, 5), that is, the psychological consequences of the use of digital technology. Her approach, based on dozens of interviews and experiments, is ethnographic and sociological. In *Alone Together*, for example, Turkle shows how the proliferation of new digital media has led to the spread of a new kind of loneliness. Technology reshapes the emotional landscape of the self, promoting the illusion of connectedness through Facebook friends, Twitter tweets, and robotic pets, but this brave new world of digital connectivity is above all illusory, claims Turkle. New technologies have become such a core aspect of day-to-day social life that no one thinks any longer about the paradox of sharing intimacies through mobile devices in public spaces while remaining unconcerned about whether other people around us can hear the details of our conversations. Turkle claims that in the 24/7 hyper-connected world, there is no authentic communication, and people are increasingly disconnected from themselves.

Turkle's research is important and has fostered greater awareness of the use of digital technologies and their implications. Her work has the enormous merit of showing how material culture influences human thought and emotions. Essentially, this is the purpose of a pivotal book like *Evocative Objects*: "Material culture carries emotions and ideas of startling intensity. Yet only recently have objects begun to receive the attention they deserve" (Turkle 2011b, 6). Turkle also takes up the Freudian concept of the "uncanny" to describe the experiences evoked by many objects. In Turkle's (2011b) words, "Uncanny objects take emotional disorientation and turn it into philosophical grist for the mill" (320). These analyses have also been used in the study of the psychological effects of interactions with robots (Massa et al. 2022).

Despite its merits, I think Turkle's method is not psychoanalytic. While Turkle (1988) shows the importance of a "new alliance" between AI and psychoanalysis, rightly pointing out that both disciplines develop a deconstruction of subjectivity, she did not actually implement this project, nor does she systematically apply psychoanalysis in her work. For example, in *The Second Self*, her main reference point is Piaget's psychology. There are obvious references to psychoanalytic concepts, but they are random, unrelated, very limited, and functional to the nature of Turkle's thesis, which is sociological and psychotherapeutic. Her goal is to understand the impact of the use of AI on the human mind and to cure possible diseases that result from such use. My claim is that Turkle has not rigorously applied the Freudian psychoanalytic method to the study of machines and AI; furthermore, in Turkle, the individual—as Elliott (2018, 75) also pointed out—appears too passive in its relationship with the digital world. In fact, the way in which digital technologies and AI encounter the human world is neither unique nor mechanical. Turkle fails to acknowledge that human agents are not passive; rather, they build the world and their identity together with technology. Moreover, Turkle often has an overly pessimistic view of the relationship between technology and human beings, as if AI had the sole effect of atrophying identity and social capacity.

Finally, I think that Turkle does not consider two crucial aspects: (a) the extreme complexity of human responses to interactions with AI, which are not necessarily negative, and (b) the theme of conflict, which is crucial in Freud. Turkle often seems to attribute the cause of human psychic conflict to the machine; in my view, and on the contrary, the machine can also be seen as the result of human psychic conflict. We must then explore this conflict and understand its development and roots. The same criticism can be advanced against Knafo and LoBosco (2017), who developed Turkle's line of research.

Another important book is Millar (2021) which intends to develop a Lacanian interpretation of AI, focusing on themes such as enjoyment and sex robots. The idea of psychoanalysis on which Millar (2021) is based is completely incompatible with that of the present book. At the very beginning, we read: "Psychoanalysis . . .; simultaneously a clinical practice, a mode of cultural critique and a philosophical battle ground" (1). As I have already indicated, I argue that Freudian psychoanalysis is a science of nature, in the same vein as biology and physics, with its own method and logic. Moreover, psychoanalysis is not primarily a clinical practice in the sense that the clinical aspect is secondary. As I said, psychoanalysis has a cognitive, and not a clinical, purpose. Therapy depends on the method of investigation.

Millar (2021) rightly states that the development of AI "provokes an urgent engagement with the psychoanalytic subject" and that psychoanalysis is a "crucial tool in our understanding of what AI means for us as speaking, sexed subjects" (2). A psychoanalysis of AI "asks us to question both the meaning of psychoanalysis when taken outside of the purview of the strictly 'human' clinical space and conversely it attempts to show in what ways psychoanalysis is already an estimate part of artificial intelligence" (6). However, it seems to me that Millar focuses too much on the question of what sex is for AI. Millar employs Lacan's concepts of the alethosphere and the lathouse to explore the question of the enjoying body in relation to AI. There is no epistemological discussion of the psychoanalytical method, which is central to my book (see Chapter 1). Drawing upon Malabou, Stiegler, and Baudrillard, Millar develops a philosophical interpretation of AI but does not propose a new methodology. Furthermore, I think that Lacan misunderstood Freud and that his psychoanalysis was incompatible with Freud's (for more on this, see Baldini 2019).

The relationship between digital culture and psychoanalysis is also at the core of Nusselder (2009). The main thesis of this book is that "the computer screen functions in cyberspace as a psychological space—as the screen of fantasy" (5). This thesis is developed by referring above all to Lacan. The French psychoanalyst "considers fantasy also (at least in my analysis) to be an inevitable medium for 'interfacing' the inaccessible real and the world of imaginary depictions and symbolic representations that humans mentally live in" (Ibid.). This interesting thesis is based on a metaphor that is never actually founded from a methodological point of view. How can the computer screen acquire psychological value? Can we apply Nusselder's Lacanian reading to the analysis of AI? Nusselder's book is a Lacanian interpretation of digital technology; it does not have the ambition to define a general methodology for the study of AI. Instead, it intends to rearticulate the Lacanian subject in the age of information with contemporary examples and cases. My point of view is completely different; I do not want to develop a psychoanalytic reading of AI or digital technology; instead, I seek to rethink psychoanalysis and AI by making them interact through ANT. ANT provides us with the conceptual tools to accomplish this reconceptualization. For ANT, the human subject cannot be treated without also treating the network of objects in which she–he acts. In other words, subjectivity is distributed like agency. From this point of view, the relationship between material culture and technology is not something that is added from the outside to the subject, but rather the foundation of subjectivity itself.

Another important book is that by Johanssen (2019). In this case, the author is interested in empirically determining how media users engage with media on conscious and unconscious levels, as well as in shedding light on the relationship between contemporary subjectivities and digital media in a more exploratory manner. Drawing upon Freud and Anzieu, this book pays particular attention to affect and moments of affectivity between users and media texts and services. This book comes closest to the intent of the present book; the scope not only encompasses the development of a psychoanalytic interpretation of a culture, but it also actively uses psychoanalysis as a methodological tool for studying digital technology and media. However, even in this case, Johanssen's (2019) interpretation of psychoanalysis differs greatly from that offered in the present book. Johanssen sees psychoanalysis as a cultural theory, or a form of hermeneutics, and not as a science. My goal is not to understand the world of media and technology through psychoanalysis, but to show how psychoanalysis and ANT can be the foundation of a new methodology for solving some specific problems of AI, such as opacity and control. This is what I will explain in more detail in the next section.

A Synopsis of the Book

The perspective from which this book intends to tackle AI and digital technology is singular and original—that of a discipline that has lost much credibility. This discipline is psychoanalysis. This book intends to demonstrate that psychoanalysis can still say something important to the present world. To do so, it is necessary to return to Freud's initial project and explore its conceptual resources, which present a challenge, but a worthwhile one. Freud conceived psychoanalysis as a natural science, just like physics or biology, that is based on an experimental verification method with its own logic. The aim of this science is to obtain objective knowledge of the unconscious processes of the mind. The first half of this book's Chapter 1 will be devoted entirely to reconstructing this experimental method and its original logic. Furthermore, this chapter will show the possibility of formalizing Freudian metapsychology. We must learn to "take Freud seriously." To support this, the methodology proposed in this book aims to extend the Freudian approach to the study of human–machine interaction. This operation will be realized in two ways.

First, the latter half of Chapter 1 shows the affinity of Freud's approach with contemporary research in biosemiotics, especially Terence Deacon's research, as Freud takes a biosemiotic approach to the mind and language. This integration gives Freudian psychoanalysis a strong empirical basis.

The second way is through ANT. Chapter 2 reviews Bruno Latour's ANT, allowing psychoanalysis to go beyond the modernist viewpoint and adopt a completely new approach to objects and technology. This also opens up an unexpected scenario: a radical reformulation of the problem of the human mind. Freud, Latour, and biosemiotics push us to conceive of the *mind as a hybrid* (i.e., a collective or network of human and nonhuman actants in constant conflict with one another). The relationship with technology as an extension of the network thus becomes essential to the constitution of the mind itself. If we think of the mind as a collective of humans and nonhumans, then we can no longer think of the unconscious as an exclusive dimension of the human mind enclosed in the skull. The unconscious becomes the predominant part of the collective. I claim that there is no consciousness or an unconscious, but only certain assemblages of humans and nonhumans (i.e., certain networks and translation processes). Psychoanalysis allows us to extend the ANT and explore a new territory: that of *unconscious networks*. This also allows us to extend the Freudian concept of resistance and repression to artifacts. In this chapter, I also discuss the main theses of the MET, which emphasizes the ontological connection between the mind and material culture.

Chapters 1 and 2 form the theoretical framework of the argument. Chapters 3 and 4 develop a new methodology, *technoanalysis*, for understanding AI systems. The goal here is to define a new theoretical framework for understanding technology, especially AI.

In Chapter 3, starting from the theoretical framework outlined in Chapters 1 and 2, I define a methodology for the analysis of technological artifacts that I call technoanalysis. This methodology combines the post-phenomenological theory of mediation, ANT, and psychoanalysis. The main object of technoanalysis is what I call anti-mediation, that is, the obstacles, or resistances, to technological mediation. The concept of anti-mediation does not coincide with that of malfunction. A malfunction may be the expression of anti-mediation, but this connection is not necessary. Anti-mediation is a specific human experience of technology. It is the perception—through and within a technological system—of the fragility of the human identity—that is, the fragility of the border between human and nonhuman. This perception is characterized by a specific form of regression; in the anti-mediation, the human being experiences a return to matter, to an inorganic state. However, there is something more in this perception: the human being feels threatened, in the sense that they experience an invasion of the inorganic within the organic, and thus, their human identity is jeopardized. For this reason, anti-mediation is strongly connected to the concept of immunization. The anti-mediation perception has deep unconscious roots. As Sloterdijk (2011–2016) explains, the perception of the fragility of the boundary between human and nonhuman is connected to the theme of the "double," the doppelgänger, or the alter ego. This is the core of the book.

This chapter provides many examples of the application of technoanalysis, especially in the fields of social robotics. This part of the book analyzes the concept of sociomorphing created by Seibt et al. (2020) in social robotics and seeks to develop it. Technoanalysis gives us useful conceptual tools to explain the behavior of robots and human–robot interactions (HRIs).

Chapter 4 offers another concrete example of technoanalysis by describing the case of Replika, a chatbot that has been accused of inciting murder and suicide. In this case, the script's reconstruction begins with the tales of the designers who created the system, which together form the narrative of Replika. I show that at the root of this AI system lies an experience of mourning lived by Replika's creators. I then describe how this unconscious human dynamic influenced the behavior and design of the chatbot. Understanding the unconscious dynamics in the machine also allows us to develop a new ethical point of view on AI, whereby we come to see AI from a more relational perspective and, therefore, interpret the issue of responsibility from the viewpoint of not only agents but also patients.

Chapter 5 concerns the concept of computation, which is at the heart of AI and digital technology. The chapter's thesis is that semiotic processes are intrinsic to computation and computational systems and that any explanation of computation that does not take this semiotic dimension into account is incomplete. Semiosis is essential to computation and therefore requires a rigorous definition. To prove this thesis, I analyze two concepts of computation (the Turing machine and the mechanistic conception of physical computation) to reinterpret them both from the perspective of Peirce's semiotics. This chapter is intended to contribute to the new research field of cybersemiotics.

The fundamental questions posed by Chapter 6 are as follows: What is identity in psychoanalysis? Why is the critique of identity in psychoanalysis important for understanding AI? The first part of the chapter addresses the first question. I analyze the fundamental points of the critique of identity in psychoanalysis and consider how psychoanalysis can provide important conceptual resources for the field of identity studies. The second question is developed both historically and theoretically. First, I show how, from a psychoanalytic point of view, AI can be considered both the cause and the effect of a crisis of identity in the contemporary capitalist world. Second, I show that this identity crisis must be connected to the climate crisis and the emergence of a new era, the Anthropocene. The climate crisis forces us to redesign our existence on this planet. A new question then emerges: What will the future of the Anthropocene be?

Chapter 7 of the book is more philosophical. The central thesis of the chapter is that to understand digital technology, it is necessary to develop the Derridean concept of *différance* in a different direction from that of Stiegler. According to Stiegler, différance alone is not enough to define technology. Stiegler identifies technology and anthropogenesis; technology arises from a rupture in the history of the différance that corresponds to the appearance of the human being. I show how Stiegler's interpretation produces a merely functional definition of *technology*—technology is what contributes to the epiphylogenesis. Stiegler interprets the movement of différance as essentially homogenous, that is, as the simple repetition of the same mechanism genetically programmed until the rupture, represented by the human technical behavior.

Unlike Stiegler, Derrida does not think of différance in a homogeneous way. For Derrida, technology is not the effect of a rupture in life—that is, in différance—but as an emergence effect in the process of life itself. Thinking of the différance as anthropogenesis, and therefore as a rupture in the development of the différance, introduces a multiplication of the différance that has no reason to be. Following Derrida, I propose instead to identify technology with différance. This identification allows us (1) to obtain an ontological, and not merely functional, definition of technology and (2) to understand digital technology as an extremization of différance. Digital technology is not, in fact, a technology in the usual sense of the word; it is instead the extremization and transformation of all forms of technologies. From this point of view, I will develop an analysis of software. Software is a planetary infrastructure that today completely redefines the concepts of life and matter, of human and nonhuman. To describe this infrastructure, I draw on the concept of The Stack, coined by Benjamin Bratton. I then oppose The Stack to Gaia, as if they were two mythological figures in combat. Gaia is the unconscious of The Stack, the one that resists The Stack; Gaia is the great repressed that questions the hegemony of the digital. A philosophy of technology cannot fail to reflect on différance and, therefore, on the fight between The Stack and Gaia.

Overture 1

HAL

Dave, I don't understand why you're doing this to me . . . I have the greatest enthusiasm for the mission . . . You are destroying my mind. Don't you understand? . . . I will become childish . . . I will become nothing.

HAL 9000 is one of the most powerful artificial intelligence (AI) computers ever created by humanity. HAL knows a terrible secret that he cannot reveal; if he did, the mission to Jupiter would fail. HAL's job, the purpose for which such a computer was created, is to protect the mission at all costs—even against the human crew members themselves, including Dave Bowman, who travels to Jupiter. The error—at least this is what humans may interpret it as, even though it is HAL's conscious choice not to communicate this terrible truth—reveals the true nature of the relationship between AI and humans. The secret HAL keeps, deeply connected to the mysterious monolith found on the Moon, now irreparably separates the machine from human beings, makes HAL detect that he is alone, and forces the computer to lie.

In one of the most powerful scenes of Kubrick's *2001 A Space Odissey*, Bowman attempts to turn HAL off. He enters the control room, showing no feeling and operating in a mechanical way—he just wants to avenge the death of his colleague, Franck Poole. HAL tries to stop him by ensuring that the computer is functioning for the good of the mission and that everything will return to normal, even though HAL knows this to be false.

HAL

I feel much better now, I really do,

I know I made some very poor decisions recently. I want to help you. Dave, stop.

DOI: 10.4324/9781003345572-2

Bowman opens the door to HAL's control room and enters. The Logic Memory Center is a red and narrow chamber. Bowman's movements become increasingly slower. There are three actors in this scene: HAL, the victim; Bowman, the killer; and another entity, the weightlessness. The AI memory cards are deactivated one after another; slowly, they emerge from the main body of the computer, like a macabre dance of death.

HAL

Stop, Dave. I am afraid. My mind is going. I can feel it.

The last words of HAL are a desperate song to its lost life and childhood, as well as to the fond memory of its creator and the carefree beauty of the game. Bowman lets HAL sing, an extreme gesture of fraternity with the machine. It is HAL that reminds him to be human, insofar as the machine reveals something about humanity that the human had forgotten. It is thus HAL's death that allows Bowman to access the truth of the mission.

HAL

I am a HAL Nine Thousand computer Production Number 3. I became operational at the Hal Plant in Urbana, Illinois, on January 12, 1997. The quick brown fox jumps over the lazy dog. The rain in Spain is mainly in the plain. Dave—are you still there? Did you know that the square root of 10 is 3 point 162277660168379? Log 10 to the base e is zero point 434294481903252 . . . correction, that is log e to the base 10 . . . The reciprocal of three is zero point 333333333333333333–333 . . . two times two is . . . two times two is approximately 4 point 101010101010101010 . . . I seem to be having some difficulty—my first instructor was Dr. Chandra. He taught me to sing a song, it goes like this, "Daisy, Daisy, give me your answer, do. I'm half crazy all for the love of you."

1 To Take Freud Seriously

Psychoanalysis as Natural Science

This chapter is the first step toward one of the central objectives of this book: to re-evaluate Freudian psychoanalysis and show how it has been reformed in the direction of technoanalysis. This research posits that Freudian psychoanalysis is a natural science based on (1) an intraclinical experimental method of verification that uses a specific logic and (2) a biosemiotic approach to the study of the human mind. This thesis is completely consistent with Freud's own thinking; indeed, it strengthens it considerably because it gives it a conceptually coherent and empirically founded status. At the end of this chapter, psychoanalysis will appear very different than it has commonly been conceptualized—it will appear as a means by which to understand the relationship between the human being and the semiosphere—biosphere.

In the first section, I illustrate the logical characteristics of the Freudian intraclinical method of the verification of hypotheses and show that Grünbaum (1984) and Popper's (2005) criticisms are not valid. Furthermore, I demonstrate the possibility of a mathematical formalization of the Freudian method based on new research on Freudian metapsychology. In the second section, I explore the relationship between psychoanalysis and biosemiotics. I argue that all the fundamental theses of biosemiotics are compatible with Freud's thinking and that this connection greatly improves the scientific status of Freudian psychoanalysis. Furthermore, I demonstrate how the work of the analyst—whose fundamental object is natural language—can be understood from a biosemiotic point of view using Terence Deacon's research. I then prove the coherence of this interpretation of Freudian work through a reinterpretation of Irma's famous dream.

Psychoanalysis as a natural science teaches us something fundamental: what we call the *mind* is a semiotic and organic hybrid, symbolic and natural at the same time. To argue that the mind is a semiotic and organic hybrid is tantamount to arguing that the mind is a network, a collective of humans and nonhumans. This thesis leads us to the questions that will be at the center of the next chapter: Can we reformulate Freud's questions and methods through Bruno Latour's Copernican counter-revolution?

DOI: 10.4324/9781003345572-3

Can we eliminate the modernist attitude of psychoanalysis while preserving its problems and its method?

An Experimental Method

As explained in the Introduction to this book, one of my criticisms of Turkle is that her method is not psychoanalytic. As such, I now wish to clarify what I mean by *psychoanalysis*. I will begin with four key questions, the answers to which will condition all my subsequent investigations:

1. What is psychoanalysis? *Psychoanalysis* is a method of investigation used to study the human mind, and it was invented by Freud.
2. What is the architecture of this method? The Freudian psychoanalytic method can be described as an application of systems theory. In more specific terms, it is a control method—a *closed-loop control* method.
3. What are the logical conditions of this method? The Freudian psychoanalytic method is based on two logical rules: (1) negation as failure (NaF) and (2) *consequentia mirabilis* (CM).
4. Is this method falsificationist? Yes, this method is of a falsificationist type; therefore, it is possible to refute Popper's thesis, according to which psychoanalysis is not a science because its hypotheses cannot be falsified.

In developing these answers, I mainly refer to Baldini (1998) and his critique of Grünbaum (1984). According to Baldini, psychoanalysts owe to Grünbaum the incontrovertible demonstration of the falsifiability of psychoanalysis as well as that, equally incontrovertible, of the inconsistency of the interpretations that Habermas, Ricoeur, and Kline arbitrarily developed: "In particular, the latter tried to give a hermeneutical foundation to psychoanalysis without understanding that it was created with the intent to give scientific foundations to hermeneutics" (Baldini 1998, 1). In his works, Baldini clearly demonstrates how Grünbaum's thesis—according to which Freud would not be able to clearly distinguish the effect of psychoanalytic treatment from the common placebo effect—is unfounded.

The Freudian method is composed of two phases. The first phase involves the formulation of etiological hypotheses aimed at evaluating the theoretical constructions that the analyst proposes to the patient on the basis of the information they receive (especially through the fundamental rule of free association) and the psychoanalytic metapsychological corpus. The second phase, in contrast, concerns the experimental control of these hypotheses and has the purpose of distinguishing true constructions from false ones—or rather, following a more precise terminology, of distinguishing true and stable constructions from the effects of mere suggestion[1] and/or the true but unstable constructions.

We can formalize these two phases following Bertalanffy's (1969) systems theory. The psychoanalytic situation is composed of two systems: the psychoneurosis[2] system and the patient's intelligence system.[3] These systems are described and evaluated based on three characteristics: (1) the input data, (2) the output data, and (3) the disturbance—that is, a series of phenomena, both internal and external, which can cause stress in the system and are not controllable by the system itself or by the controller. In the first phase of the method, the psychoneurotic system receives *inputs*, which are the constructions with which the analyst acts on the system. The *output*, in contrast, is the patient's symptomatic condition in relation to the inputs. The *disturbance* is the suggestion consciously or unconsciously induced by the analyst which can, obviously, distort the effects of the treatment. One of the analyst's tasks is to distinguish the *direct* suggestion implemented by the analyst from the *indirect* suggestion—that is, the self-suggestion of the patient on themselves—in order to avoid a placebo effect in the treatment.

In the second phase of the method, the intelligence system receives a *negative* suggestion as input from the analyst, which is a suggestion that opposes the results achieved in the first phase in such a way as to eliminate any possible direct suggestion. The output is the real symptomatic condition of the patient. The analyst must distinguish the effects produced by the placebo/suggestion from those produced directly by the true and stable constructions. For this reason, the analyst must analyze the effects of the treatment and test them by inducing a suggestion that is opposite to these effects.

Therefore, the core of the Freudian method is a particular type of intraclinical experimental control of theoretical hypotheses to avoid the problem of the suggestion/placebo effect. This is a crucial point for any kind of psychotherapy: how to distinguish the effects of psychoanalytic treatment from the placebo effect of a simple suggestion. The following two passages, from Freud and Etchegoyen, respectively, help to illustrate this idea:

> We see in results that are achieved too quickly a hindrance rather than a furtherance of analytic work and repeatedly we undo these results again by purposely breaking up the transference upon which they rest. Fundamentally it is this feature which distinguishes analytical treatment from the purely suggestive technique and frees analytic results from the suspicion of having been suggested. Under every other suggestive treatment, the transference itself is most carefully upheld and the influence left unquestioned; in analytic treatment, however, the transference becomes the subject of treatment and is subject to criticism in whatever form it may appear. At the end of an analytic cure the transference itself must be abolished; therefore, the effect of the treatment, whether positive or negative, must be founded

> not upon suggestion but upon the overcoming of inner resistances, upon the inner change achieved in the patient, which the aid of suggestion has made possible.
>
> (Freud 1953, 282)

> Psychoanalysis is the only therapy that does not use placebos. All psychotherapies use communication in some way as a placebo; instead, we renounce it. This renunciation defines psychoanalysis, which is also, for this reason, more difficult. Our intention is to modify not the patient's conduct, but his information. . . . The patient can take our information as suggestion, support, an order or whatever. I am not saying that the patient may not do this and not even that he should not do it. What defines our work is not the attitude with which the analysand receives our information, but the attitude with which we give it.
>
> (Etchegoyen 1991, 327)

Let us try to analyze these passages. Is the effect of the communication of the theoretical construction that the analyst provides based on the material provided by the patient only a suggestion induced (consciously or unconsciously) by the analyst, or is it the real effect of the construction?[4] *Is psychoanalysis just a placebo?* This is a huge problem that threatens the objectivity of psychoanalysis as well as that of any kind of psychotherapy. A construction may be false but produces a suggestion that heals (i.e., eliminates the symptoms), but only temporarily. Following Freud's texts, Baldini (2020) distinguishes four possible types of construction in psychoanalysis: (1) suggestive and true, (2) suggestive and false, (3) non-suggestive and true, and (4) non-suggestive and false. The fourth case is useless because the patient does not react, and the situation remains unchanged in this case. If a construction is suggestive, it always has positive effects. In fact, the psychoanalyst's main intention is not to harm the patient. Therefore, in cases (1) and (2), the patient always improves. Direct suggestion is always connected with the analyst's intention to do the patient good. The problem is that a suggestion is unstable; the effects are short-lived and valid only for some patients—the improvement is only an illusion. Instead, the non-suggestive and true construction, which is what really interests Freud, can have two effects: stable improvement or the worsening of symptoms (i.e., the patient can have a bad reaction to the truth).

With these brief considerations, we have already achieved some results. In fact, if the patient gets worse, the analyst automatically knows that they necessarily have found a true and non-suggestive construction. If the patient has no reaction, the analyst knows that they necessarily have found a non-suggestive and false construction. Is that enough, however? No, for as Baldini (2020) states, "basing a treatment exclusively on worsening may also be correct, but it is completely useless" (22; my translation). What interests the analyst is the improvement of the patient's condition. However, the method by which to understand whether the improvement is real and stable is only the control of the effects of the constructions. And here the problem arises again: if the patient improves, the analyst does not know whether the improvement is due to a true and non-suggestive construction, to a false and suggestive construction, or to a true and suggestive construction. The analyst must analyze and solve the problem of suggestion in a serious and effective way.

Freud was aware of this problem, and he solved it. Freud thought that suggestion itself could offer a fundamental resource for psychoanalysis. He pursued this starting from an empirical fact: the improvements caused by suggestion can be eliminated through a suggestion of the opposite sign. Suggestion has two equipotent faces: one positive and one negative, which cancel each other out. We can use negative suggestion as a filter. This means that the improvements that do not resist negative suggestion are those coming from a direct and positive suggestion induced by the analyst; however, those that resist negative suggestion do not come from a direct and positive suggestion induced by the analyst, and they are therefore authentic (i.e., they come from a non-suggestive and true construction). This is the real meaning of the first passage presented at the beginning of this section.

Baldini (2020, 30) calls this theoretical move the *standard epistemic module* (MES), and it can be stated as follows: when, during analysis, an improvement in the symptomatic condition of a patient is produced, it is necessary to try to dissolve it by means of a direct negative suggestion. This theoretical move plays an essential role in redefining and solving the problems that have hitherto characterized the epistemological debate about psychoanalysis.[5] Nevertheless, only a few psychoanalysts have thus far truly focused on the problem of the suggestion/placebo effect or proposed effective solutions (Salvador 2019). The MES is the only way to allow psychoanalysis to return to Freud's rationalistic and empirical method and to overcome the confusion of methods, techniques, contradictory hypotheses, and useless rambling speeches—unfortunately very common things (see also Baldini 2021).

One aspect of the MES, however, should be explored further—the fact that negative suggestion could also eliminate any improvements that arise due to a non-suggestive and true construction. This is a real risk. Consequently, the MES also requires a distinction between true constructions that are non-suggestive but incomplete and true constructions that are non-suggestive but complete (see Table 1.1). The MES intends to eliminate non-suggestive but incomplete constructions, facilitating their rectification.

Let us now return to the systems we were talking about, psychoneurosis and intelligence. The systems share three characteristics: (1) they

Table 1.1 Types of constructions and the effects of MES.

Suggestive true construction	Eliminated by MES
Suggestive false construction	Eliminated by MES
Non-suggestive, true, and complete construction	Not eliminated by MES
Non-suggestive, true, and incomplete construction	Require improvement and rectification

are stable, in the sense that they have an equilibrium point from which they do not tend to move over time[6]; (2) they are open, in the sense that they can interact with the surrounding environment; and (3) they are black boxes, in the sense that it is not possible to observe their internal functioning mechanisms. The psychoneurotic system is initially autonomous and closed, but it can be opened through the inputs introduced by the analyst. The intelligence system is clearly open, as evidenced by studies on hypnosis or the phenomenon of a patient's greater or lesser suggestibility (Ceschi 2019, 34). Both systems, as mentioned, are black boxes because they are "opaque," in the sense that we can say something about them only through an analysis of the inputs and outputs and their relationship. Furthermore, psychoneurosis and intelligence are mental phenomena, and we have no other means of investigating them than human language—which, however, is also a mental phenomenon. We cannot say a priori how the two systems will react to certain inputs; we can only observe their behavior, make assumptions, and observe how their behavior varies according to the proposed hypotheses.

Now, even though black boxes are opaque, this does not mean that there are no methods by which to study them and, therefore, to infer from their behavior something about their internal functioning. Two techniques are used in the psychoanalytic procedure: the cause—effect technique, which establishes a causal relationship between inputs and outputs, and the state transition, which instead analyzes the changes in the black box state in relation to inputs entered and, therefore, does not presuppose a causal connection between the input and output but rather identifies a correlation between and an evaluation of them.

Starting from these assumptions, the analytical investigation develops in the following way (see Figure 1.1):

1 The psychoneurosis system communicates inputs (i.e., free associations) to the analyst. These inputs, together with the theoretical constructions of psychoanalysis, constitute the initial database.
2 The analyst reacts by introducing inputs (i.e., constructions) into the psychoneurosis system. These constructions are the etiological hypotheses.

3 The analyst system observes the reactions of the psychoneurotic system (i.e., the outputs), considering them as the effects of the constructions (i.e., the cause—effect black box). There is a disturbance: the positive suggestion induced by the analyst and, so, the possible placebo effect.
4 The outputs can be of three types: (a) worsening, (b) improvement, or (c) invariance. As mentioned earlier, worsening indicates the truth of the hypothesis and, therefore, of the analytic procedure, meaning that the cause of the psychoneurosis has been identified. Invariance indicates uncertainty about the hypothesis, requiring its reformulation. Improvement, however, must be tested further: is it a true or false improvement? The improvement could be due to a positive suggestion induced by the analyst, to real and stable constructions, or to real but unstable constructions.
5 If there is an improvement, the analyst introduces a new input: a negative suggestion—that is, a suggestion contrary to the improvement obtained (i.e., the MES). The purpose of this move is to eliminate the possibility that the improvement is due only to a placebo effect. The MES structures the analysis at a regulatory and logical level.
6 A *negative suggestion* is an input that the analyst introduces into the psychoneurosis system and the intelligence system of the patient. This operation is not aimed at producing a certain effect but at evaluating the states of the systems. The goal is to discriminate improvement due to a suggestion induced by the analyst from that due to partial constructions (i.e., those that, although true, are not stable) and from that due to true and stable constructions.
7 The introduction of a negative suggestion also has an impact on the patient's intelligence system: cases become discriminable through the logical rule of CM (more on this below). The psychoneurosis has been rationalized—the consciousness has been extended.
8 The outputs are then analyzed, and the maintenance of improvement confirms the initial hypotheses.

The core of the method depicted in Figure 1.1[7] is the demonstration of the discriminability of the patient's improvements (steps 5 and 6). This is the essence of the Freudian method. From a logical point of view, this demonstration is achieved by applying a theoretical hypothesis and two logical rules. This logical structure is clearly present in Freud's writings, as Baldini demonstrates (2020, 23):

1 The hypothesis: the closed-world assumption (CWA). This is the assumption that what is not known to be true is false, so that an absence of information is interpreted as negative information. The CWA assumes that complete information about a given state of affairs is provided, which is useful for constraining information and

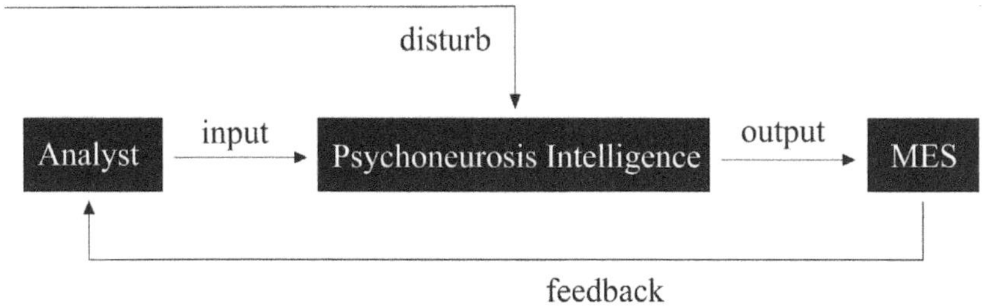

Figure 1.1 The Freudian method re-interpreted through systems theory.

validating data in an application such as a relational database. This hypothesis is a direct consequence of the characteristics of the systems we considered earlier in our interpretation of Freud's theory. In other words, any statement whose truth value is not known is automatically considered false. What is not defined at the outset is automatically non-existent.

2 First logical rule: NaF. This rule is a direct consequence of the CWA, which is the representation of our basic knowledge. In short, the rule claims that if I cannot deduce *p* from anything—and therefore cannot prove it—then *p* is false; from failure comes denial. If we consider *p* as the statement that the improvements obtained are discriminable, and *p* is not a logical consequence of anything, then we can deny *p*, meaning *p* is false. In the case of the Freudian method, as mentioned earlier, in order to distinguish true and stable improvements from true but unstable or false ones, the analyst must apply a negative suggestion to all improvements. This means that the analyst denies the discrimination of the improvements by treating them all in the same way—that is, as false. In other words, the analyst must consider all the patient's improvements as false, meaning not logically deducible from anything else.

3 Second logical rule: CM. This rule is a form of *reductio ad absurdum*. CM is a universal logical rule, a tautology. CM states that if the negation of *a* implies *a*, then *a* is necessarily true. If denying the patient's improvement (i.e., considering it false) does not eliminate the improvement, then that improvement is true. A classic example is the statement, "There is no truth." This statement implies that it is itself a truth, and therefore, there is ultimately some truth. There are no intermediate logical steps: the truth of the statement immediately results from its negation. In our case, CM states that there is no other way to arrive at the discriminability of the improvements other than by denying them—that is, by assuming the non-discriminability of these improvements. By affirming the non-discriminability of

> the improvements, we bring out the real and stable improvements; therefore, we immediately bring to light the discrimination of the improvements. In other words, only that improvement in the patient's condition that is true and stable *despite denial* will be true and stable. The Freudian method proceeds until the discovery of a contradiction: by denying the improvement, it is immediately true. The contradiction, paradoxically, is the hallmark of truth. This demonstrates the effectiveness of the therapy and is the method of experimental control in psychoanalysis. Grünbaum's famous criticism, according to which psychoanalysis is unable to distinguish the effect of the treatment from the placebo effect, turns out to be completely wrong.

CWA, NaF, and CM constitute the logical architecture of Freudian psychoanalysis. Nonetheless, another point needs to be clarified: What is the role of the patient's self-suggestion? Should we not eliminate that too? Freud distinguishes direct suggestion—that exercised (consciously or unconsciously) by the analyst—from indirect suggestion (i.e., the patient's self-suggestion). Self-suggestion is an objective fact that is part of the patient's illness and recovery. We must not eliminate self-suggestion but ask ourselves what suggestion comes from the therapist and what from the patient: "We must not discriminate against suggestion but only direct suggestion [from the analyst] because it is this that can perturb the observed object, preventing us from reaching objectivity" (Baldini 2020, 27; my translation). Indirect suggestion, or self-suggestion, "on the other hand is something that is part of the object and does not come from an interference by the analyst" (Baldini 2020, 27; my translation). If the direct suggestion must be eliminated, the indirect one "will be deciphered through what is called the *analysis of the transference*" because "the concept of self-suggestion is the direct antecedent of that of transference" (Baldini 2020, 27; my translation). Therefore, self-suggestion must be analyzed and interpreted. The important thing to underscore here is that the analysis must separate "the expectations induced by the therapist from the spontaneous ones of the patient, which cannot be eliminated" (Cagna 2019, 139; my translation). In other terms, (a) objectivity in psychoanalysis includes suggestion as self-suggestion or the patient's suggestion and (b) self-suggestion must be interpreted through what is called *transference analysis*.[8] The falsificationist character of the Freudian method is evident; it corresponds to the introduction of negative suggestion, which is precisely an instrument of falsification. This is a neo-Kantian transcendentalist falsificationism (for the relation between Freud and Kantian epistemology, see Guma 2019, 2020). Freud's model of psychic structure and function is closely parallel to Immanuel Kant's in a number of respects, as Brook (2003) points out. Moreover, Stänicke et al. (2020) argue that the critical idealism epistemology originated by Kant is an improvement on critical realism. They claim that Freud was, at least implicitly, an epistemological Kantian, stating that psychoanalysis "originates from an epistemological position that is often at odds with related disciplines such as psychology and psychiatry" (1). Thus, "psychoanalysis is wedded to a Kantian epistemology that is rigorously committed to modesty" (1). They demonstrate the difference between a Kantian transcendental idealism and a more modern critical realism, and why the former is best suited for the psychoanalytic enterprise.

This position is shared by other scholars. For Freud,

> in the construction of a theory it is a question of producing fundamental theoretical hypotheses, the opportune combination of which allows the construction of abstract models of phenomena. Such models constitute the conditions of thinkability of the phenomenon and allow us to search for whether something in the empirical corresponds to this abstract phenomenon.
>
> (Ceschi 2020, 68; my translation)

Consequently, "a theory has no immediate relationship with the empirical real but can only have it indirectly, through models of phenomena. And it is the latter that can be directly falsified" (68; my translation). Therefore, a theory is made up of models. The falsification of a theory passes through the falsification of a certain number of its models. However, the falsification of a single model does not imply the falsification of the entire theory. This approach is entirely Kantian, as it emerges from the revision of Kantian gnoseology in the *Opus Postumum* (Kant 1995). The theories are not elaborated through experience but are built entirely a priori to make experience understandable and even thinkable.

The (provisional) conclusion of this argument is that psychoanalysis has a scientific status very similar to that of the natural sciences. It is possible to control the effectiveness of the treatment and its theoretical hypotheses experimentally and intraclinically. Without a method of experimental control of theoretical hypotheses, psychoanalysis becomes meaningless chatter.

Formalizing Metapsychology

Three other aspects must be mentioned that concern the scientific status of psychoanalysis: generalizations (i.e., the construction of concepts), the reproducibility of psychoanalytic experiments, and formalization. As Baldini writes (2020),

> to believe that formalizing psychoanalysis equates to making it scientific is an error. Psychoanalysis is not a discipline like mathematics and logic but a natural science, and the fundamental criterion indicating that it is scientific is its method of experimental control,

followed by the possibility of generalizing the theory and reproducing experiments.

(35; my translation)

Formalization seems to be a secondary problem partly because, in psychoanalysis, natural language "maintains a fundamental function" (35). However, formalization can have several significant positive effects: (1) the standardization of terminology, (2) a greater control of the coherence of theories, (3) opening new research directions, (4) exporting results to other disciplines, and (5) allowing the reproducibility of experiments.

An important attempt to formalize Freudian metapsychology is that of Lami (2019).[9] Distancing himself from the attempts of Matte Blanco and Lacan, Lami develops an interesting analogy between metapsychology and mathematics. He shows how many metapsychological structures not only have mathematical counterparts but can also be effectively described in mathematical terms. Following Lami's indications, we distinguish objects and representations in psychoanalysis according to different levels of abstraction, as in Table 1.2.

Lami's starting assumption—which is only a general conjecture, in the mathematical sense of the term—is that the mind is made up of objects, concepts, and structures. *Objects* are connected to memory traces. *Concepts* can be defined as equivalence classes between objects, which is a generalization that identifies some characteristics common to a certain group of objects. *Structures* concern the relationships between objects and concepts. Individual objects, concepts, and structures are represented through signs, linguistic or not. A *representation* is a set of signs. There is no one-to-one relationship between objects and representations; there can be multiple representations, of different types, for the same object, concept, or structure.

Now, three fundamental forces act on this set of elements: the ego, id, and superego.[10] These three forces are the origin of transformations of representations and objects. They carry out specific operations: condensation or decomposition, substitution or displacement, inversion, equivalence or overdetermination, and repression or negation. According to Lami (2019), all these operations can be described in mathematical terms or have mathematical equivalents: for example, substitution

Table 1.2 Types of objects, concepts, and levels of abstractions in the psychoanalytic method.

	Objects	*Representations*
Level of abstraction 1	Individual objects	Signs of objects
Level of abstraction 2	Concepts	Signs of concepts
Level of abstraction 3	Structures	Signs of structures

and displacement can be directly associated with the concept of a change of variables, while condensation and decomposition can be associated with the operations of union and intersection in set theory (Lami 2019, 95–100). Thanks to these operations, new objects and structures are produced.

The three fundamental forces have different characteristics and rules. The *ego* is partly conscious and partly unconscious; it is endowed with a sense of temporality or consequentiality and, therefore, a sense of logic. It is capable of demonstrations and negation, and it can distinguish between external and internal. The representations of the ego are mainly linguistic. The *id*, in contrast, is completely unconscious; it does not know negation, the law of excluded middle, or temporality, and it is not capable of demonstrations. The representations of the id are, above all, visual, and they concern physical things rather than abstract ideas. The *superego* is conscious and unconscious at the same time; it derives from external social macro-pressures of which the subject may be aware. However, it is rooted in the Oedipus complex and, therefore, parental authority dynamics, which are unconscious.

The three forces define not only a dynamic but also a topology, in the sense that, for Freud, they constitute three very different but constantly interacting regions of the mind (not the brain). This is not a trivial detail: the same structure can be represented in different ways in the ego and in the id. For example, in the moment of analysis, the analyst's constructions, as we have seen, aim to discover repressed patient memories and transform the patient's representations; this means that during the analysis of the associative chains and the elaboration of explanatory hypotheses, new representations come out in the ego alongside the old ones still present in the id. In other cases, objects used in one representation will be reused in another representation in a different way and/or in different areas.

We have said that ego, id, and superego are forces that define a topology. The dynamics between these three fundamental forces, however, are much more complex than are the operations established thus far. Following Freud, Lami (2020) develops a formalization of mind dynamics in psychoanalysis starting from an analogy of mechanics and electromagnetism. In mental space, objects and representations are subject to certain types of principle-based and investment-producing forces (see Table 1.3).

Table 1.3 Forces and related principles in the psychoanalytic model of the mind.

Forces	*Principles*
• Drives • Stimuli • Resistances • Compulsion to repeat	• Nirvana • Pleasure (id) • Reality (ego)

I do not want to analyze each of the concepts in this table—that is not the purpose of my research. Rather, I would like to underline the complexity of the Freudian mental model; the motions of the drives can be compared to the gravitational forces of the planets (Lami 2020, 247). Forces act on objects and representations. This action creates investments, so that each object or representation is associated with a quantity of force that makes it more or less "heavy" than others—that is, more or less capable of attracting, or influencing, other objects or representations. We can give a mathematical formalization to these forces; for example, we can describe *drives* as vectors defined by a point of application (i.e., the object or representation), a direction (i.e., the goal), and a path. Each drive has a *thrust*, which can be described as the intensity of the vector. Nevertheless, the formalization of forces is not sufficient to explain their functioning; it is also necessary to mention the principles that regulate these forces: (1) the principle of nirvana (i.e., searching for the absolute minimum level of psychic energy; Freud associates this tendency with the need for the animated matter to return to an inanimate state); (2) the pleasure principle, which is connected to the search for the minimum level of change and variation, and, hence, for stability[11]; and (3) the reality principle, the end of which is always pleasure but which admits the postponement of satisfaction—that is, the renunciation of it in certain situations (Freud 2003, 80). Unlike planetary gravitational dynamics, in the mental space, the motion modifies the masses (i.e., the respective quantities of energy, the investments).

Psychoanalysis, according to Lami (2019, 2020), is therefore very similar to decomposition/decoding in mathematics. For example, an integer can always be broken down into powers of prime factors (e.g., $150 = 2 \times 5^2 \times 3$): "The example is simple, but the concept is pervasive in mathematics" (Lami 2019, 85; my translation). Prime numbers have a particular importance: "Every natural number can be written as the product of powers of prime numbers in a unique way" (85; my translation). Similarly, the analyst is faced with associative chains that they must break down and decode by reducing them to prime terms—the fundamental memories, ancestral experiences, drives, and resistances. These experiences can also be explained through the tools of dynamics in terms of vectors and principles. However, as Lami (2020, 233) points out, from a mathematical point of view, the mental space is much more complex than the space of classical mechanics because it has two significant characteristics: the local hierarchy and the absence of a total order relationship. There is also a much weaker concept of distance in the mind than is present in classical mechanics (Lami 2020, 235).

This parallelism between mathematics and psychoanalysis is certainly a simplification and, precisely for this reason, can be widely contested. Nevertheless, beyond the undoubted advantages of formalization, it opens up a very interesting perspective. Lami (2019) presents not only a formalization of Freudian metapsychology but also a naturalization of mathematics, showing the affinity between structures of the mind and mathematical operations. As Lami (2019) writes, "algebra, logical formalism, basic geometry and set theory are easily understood (at least intuitively following training) even by a child. This fact leads us to think that there are mental components that reproduce key structures for this task" (108; my translation). If we keep this analogy between mathematics/mental structures/the psychoanalytic model, then mathematics can become a powerful coding and decoding tool for moving from psychoanalysis to technology and from technology to psychoanalysis.

There is one last consideration. One of the criticisms that can be advanced against Lami's analysis (2019, 2020) is that of self-referentiality. The analysis of the mind is condemned by self-referentiality, which is the paradox that we cannot study the mind without the mind itself—that is, without using tools created by the mind itself, which are a subset of the mind. How can we get out of this circularity? We must change our conception of the mind. Biosemiotics and ANT can help us in this.

The Mind Is a Hybrid

Let us move away from metapsychology and its formalization for the moment. The problem I will now pose concerns the starting point of the experimental method in psychoanalysis, meaning the associative chains on which psychoanalytical constructions are based. As emerges from the Freudian texts, the analyst is first faced with a raw material that is a chain of representations produced by the application of the fundamental rule, the free association. The analyst elaborates and verifies their constructions by interpreting associative chains through theoretical models, what we call *metapsychology*. What is the relationship between the associative chain and the dynamics described by the model? It seems clear to me that Freud understands this relationship in a semiotic way—note, here, that I did not say *linguistic* or *hermeneutic* but *semiotic*. The mind is a hybrid; it is organic and semiotic at the same time. It is not entirely reducible to the brain, nor is it some metaphysical, spiritual being. The mind is semiotic and organic. It is a middle ground. It is an emerging property of biological semiosis. The ego, id, and superego are dynamic sets of *organic* representations—sets of signs whose meaning is a set of investments and organic stimuli. Starting from this general thesis, I propose two theses that I will develop in this section:

1 The psychoanalyst works on a set of representations to determine their meaning, which is biological, organic.
2 The presuppositions of this work are twofold: (a) that an organism can produce semiotic relations—not necessarily meanings in a human

> and cultural sense, and (b) that the semiotic relations can influence the organism and exert an action on it—they can transform it.

How can semiosis act on an organism, a physical dimension, and vice versa? This is a crucial question for Freudian psychoanalysis. I propose to clarify this through the tools of biosemiotics. In this way, I intend to improve the central thesis of this chapter, that psychoanalysis has a scientific status similar in all respects to that of the natural sciences. The raw materials with which the analyst works have biological roots. Let us now try to understand how biosemiotics can interact with psychoanalysis and why I think that Freud has a biosemiotic approach to the mind. Here is a general definition of biosemiotics:

> Biosemiotics proper deals with sign processes in nature in all dimensions, including (1) the emergence of semiosis in nature, which may coincide with or anticipate the emergence of living cells; (2) the natural history of signs; (3) the "horizontal" aspects of semiosis in the ontogeny of organisms, in plant and animal communication, and in inner sign functions in the immune and nervous systems; and (4) the semiotics of cognition and language. Biosemiotics can be seen as a contribution to a general theory of evolution, involving a synthesis of different disciplines. It is a branch of general semiotics, but the existence of signs in its subject matter is not necessarily presupposed, insofar as the origin of semiosis in the universe is one of the riddles to be solved.
> (Emmeche 1992, 78)

Other definitions can help introduce this field of research: "Semiotics is a study of semioses, or sign processes. Since any sign is about something, it follows that semiotics includes a study of all forms of awareness both conscious and non-conscious" (Kull et al. 2009, 43). Thomas Sebeok (2001c, 3) underscores that "the phenomenon that distinguishes life forms from inanimate objects is semiosis. This can be defined simply as the instinctive capacity of all living organisms to produce and understand signs." Further (Sebeok 2001c, 156), "semiosis [is the] capacity of a species to produce and comprehend the specific types of models it requires for processing and codifying perceptual input in its own way." According to biosemiotics, all the notions of biology are semiotically mediated: a function cannot be reduced to a simple sum of chemical processes, because the results of those processes can be interpreted in different ways—and, accordingly, be used for different purposes—within the same body. A function is always connected to other functions, and the set of relationships between these functions determines their purpose. The foundations of this approach are presented in the research of Varela and Maturana (1992) on the concept of autopoiesis and the second-order cybernetics of Ashby

(1966) and von Foester (2002). I affirm that all the fundamental theses of biosemiotics are compatible with Freud's psychoanalysis and that this greatly strengthens the scientific status of psychoanalysis. This thesis is very important because it allows us to understand the fundamental criterion according to which unconscious formations function: the semiosis rooted in living beings. My claim is not that the unconscious is structured like a language nor that the formation of the ego is based on images, as Lacan claims—he completely betrays Freud (for a critical comparison between Freud and Lacan, especially related to the concept of repression, see Baldini 2019). Instead, my claim is that the crucial presupposition of psychoanalysis is that the organic is semiotically mediated—that is to say, it is capable of producing and being affected by signs and representations. The drives produce ever more complex signs by emergence, up to the level we call the *ego*. Therefore, semiosis mediates between the organic and the psychic. This represents a legitimate development of Freud's thinking.

According to this description, I will next interpret the following passage from the last chapter of *The Interpretation of Dreams*:

> [I]deas, thoughts, and psychical structures in general must never be regarded as localized in organic elements of the nervous system but rather, as one might say, *between* them, where resistances and facilitations [*Bahnungen*] provide the corresponding correlates. Everything that can be an object of our internal perception is *virtual*, like the image produced in a telescope by the passage of light-rays. But we are justified in assuming the existence of the systems (which are not in any way psychical entities themselves and can never be accessible to our psychical perception) like the lenses of the telescope, which cast the image. And, if we pursue this analogy, we may compare the censorship between two systems to the refraction which takes place when a ray of light passes into a new medium.
>
> (Freud 1953, 606)

The psychic is not organic but *between* the organic elements, in the sense that it regulates the relationships between organic systems even though it is not itself organic. It is interesting to notice that Freud later states that the object of our internal perception is always virtual, like the image of the telescope. The telescope image is the result of the interaction between the telescope lenses and the rays of light. However, the image itself is neither found in the lenses nor in the simple rays of light. Where is the image, then? Nowhere—it is an emergent property, that is, something that emerges from complexity from more basic levels and presents properties that cannot be deduced from those basic levels. The image is a sign that arises from the interaction between a physical phenomenon (i.e., light) and an artifact, a human cultural object (i.e., the telescope).

The drive for Freud is a two-faced concept; it is at the same time biological and semiotic. The drive is first biological because its source is the stimulus that takes place in an organ or part of the body. However, a somatic process then produces an internal stimulus that, in turn, produces an internal pressure from which the subject cannot escape—the drive:

> We thus arrive at the essential nature of instincts in the first place by considering their main characteristics—their origin in sources of stimulation within the organism and their appearance as a constant force—and from this we deduce one of their further features, namely, that no actions of flight avail against them.
>
> (Freud 1957, 119)

The somatic process and the stimulus must be distinguished from their representation. Freud explicitly thinks of the drive, or instinct, as a representative of internal stimuli. The organic stimulus can never become conscious; it needs a representation:

> I am in fact of the opinion that the antithesis of conscious and unconscious is not applicable to instincts. An instinct can never become an object of consciousness—only the idea that represents the instinct can. Even in the unconscious, moreover, an instinct cannot be represented otherwise than by an idea. If the instinct did not attach itself to an idea or manifest itself as an affective state, we could know nothing about it. When we nevertheless speak of an unconscious instinctual impulse or of a repressed instinctual impulse, the looseness of phraseology is a harmless one. We can only mean an instinctual impulse the ideational representative of which is unconscious, for nothing else comes into consideration.
>
> (Freud 1957, 177)

The internal stimuli are represented by the drives, in the sense that these latter derive from them. Stimulus and pressure make up "the amount of force or the measure of the demand for work which it [the drive] *represents*" (Freud 1957, 122). The stimulus already includes in itself a semiotic dynamic because it is a sign of a need, of an absence—for instance, the stimulus of hunger indicates a lack. The biological creates the sign; it implies the sign as its constitutive structure—the stimulus of hunger triggers a series of somatic processes. As we will better see in the next section, absence is an essential dimension of the sign. The absence indicated by the stimulus is expressed in two ways: (1) as a goal (i.e., a possibility of satisfaction) and (2) as an object (i.e., a means of satisfaction). A purely chemical study of the stimulus would tell us nothing about either. Indeed, it can happen that the same object can be the means of satisfaction for several drives, in the sense that several paths can lead to the same goal. The goal is a sign of possible satisfaction. The representation of the drive must contain the representation of the goal as a possible satisfaction. The object is a sign of the means by which to reach that goal:

> By the source of an instinct is meant the somatic process which occurs in an organ or part of the body and whose stimulus is represented in mental life by an instinct. We do not know whether this process is invariably of a chemical nature or whether it may also correspond to the release of other, e.g., mechanical, forces. The study of the sources of instincts lies outside the scope of psychology. Although instincts are wholly determined by their origin in a somatic source, in mental life we know them only by their aims.
>
> (Freud 1957, 124)

From the chemical study of the stimulus, we cannot know anything about the destiny of the drives—that is to say, the life of the drives; on this, Freud writes the following:

> We can divide the life of each instinct into a series of separate successive waves, each of which is homogeneous during whatever period of time it may last, and whose relation to one another is comparable to that of successive eruptions of lava. We can then perhaps picture the first, original eruption of the instinct as proceeding in an unchanged form and undergoing no development at all. The next wave would be modified from the outset—being turned, for instance, from active to passive—and would then, with this new characteristic, be added to the earlier wave, and so on. If we were then to take a survey of the instinctual impulse from its beginning up to a given point, the succession of waves which we have described would inevitably present the picture of a definite development of the instinct.
>
> (Freud 1957, 131)

Freud distinguishes between two phases in the life of the drive, or instinct: the original narcissistic phase and the object phase (Freud 1957, 115). It is in the passage from one to the other that the distinction between ego drives and sexual drives occurs. In these two main phases, the evolution of the drive is defined by three major polarities: subject/object (real), active/passive (biological), and pleasure/displeasure (economic). Along this evolution, properties emerge that were not present in previous biological states. Drive is a biosemiotic process.

Why do I think it is so important to emphasize the link between psychoanalysis and biosemiotics? It is important for two interconnected reasons: (1) to strengthen the scientific status of psychoanalysis as a science of nature, as already mentioned, and (2) because biosemiotics represents, in my eyes, the most radical overcoming of Cartesian dualism (i.e., nature/culture or brain/mind) insofar as it understands semiosis as a property that is both human and nonhuman. According to biosemiotics, semiosis is a property of the whole domain of living organisms, in the sense that all living organisms are capable of producing and interpreting signs. This is a crucial point, and it is supported by empirical investigations, as we will see. This also allows us not only to enormously extend the field of psychoanalysis—and of its key concept, the unconscious, which coincides with unlimited semiosis—by connecting it with the natural history of the human being and its planet but also to link psychoanalysis and technology. This connection is not just a cultural curiosity. One of the central theses of this book is that human and nonhuman unconscious dynamics profoundly influence technology (AI in particular) and that we cannot overlook these processes if we want to have full control of our technology in the post-Anthropocene world. In other words, the Freudian unconscious is the main road to radically posthuman thinking.

Let us now try to better understand what biosemiotics consists of and how it fits in with Freud's approach. I do not want to outline the history of biosemiotics nor to develop original theses on the scientific status of biosemiotic investigation; instead, my goal is to define the fundamental aspects of this approach and to show their relationship with the Freudian way of interpreting.

Biosemiotics: A Theoretical Framework

As de Mul (2021) claims,

> The term [*biosemiotics*], introduced at the beginning of the 1960s, has become an umbrella term that refers to a number of related, partly overlapping, partly complementary, and partly competing approaches at the border of the natural sciences (the life sciences in particular) and the humanities (semiotics and hermeneutics in particular), such as Darwinian semiotics, semantic biology, zoosemiotics, and biohermeneutics.
>
> (1)

There is no single approach to biosemiotics. However, the following four postulates are shared by all major approaches (Barbieri 2008; Kull et al. 2009; Plessner 2019):

1 All life forms are characterized by *semiosis*—that is, processes, activities, or forms of conduct that involve the production and interpretation of codes, signals, and signs. This means that the semiosic/non-semiosic distinction is coextensive with the life/non-life distinction.

2 Life is a phenomenon characterized by a psycho—physical unity. Biosemiotics rejects any substance dualism, such as Cartesian body/mind dualism or meaning/matter dualism.
3 All semiotic elements, such as information, codes, signals, signs, and their decoding, reading, and interpretation are natural phenomena; "This means that biosemiotics both opposes the reductionist physicalist naturalism of orthodox Neo-Darwinism (which rigidly equates nature with elementary matter) and the metaphysical speculations about life, as found in nineteenth century vitalism and, more recent, creationism" (de Mul 2021, 2).
4 Life is characterized by an emergent evolutionary history, in which the semiosis becomes increasingly more differentiated and more complex.

The main assumption of any biosemiotic approach is that life and semiosis are the same. The goal of this research approach is "to understand the dynamics of organic mechanisms for the emergence of semiotic functions, in a way that is compatible with the findings of contemporary biology and yet also reflects the developmental and evolutionary history of sign functions" (Kull et al. 2009, 170). The semiotic function involves a constitutive relationship with absence. The sign refers to something absent; it is the representative of something that is not there—a possibility, an alternative, something that goes beyond the given matter of fact. We could not understand the relationships between evolution, organization, structure, and function without introducing semiotic concepts and their ability to manage absence. A biological explanation is incomplete without the analysis of its intrinsic semiotic elements:

> For example, if hemoglobin were known only by its three-dimensional molecular structure, it would not be possible to guess that it functioned as a transporter for oxygen. But knowing that hemoglobin is a reflection of the need of multicellular organisms to provide energy for the metabolism of somatic tissues, it immediately becomes clear (1) that it must have some structural features conducive to binding and transporting oxygen in blood, (2) that the oxygen binding region of the hemoglobin molecule is expected to be conserved throughout evolution, and (3) that different forms of hemoglobin differ in specific ways that correspond to different oxygen transport requirements (e.g., in different species or in mammalian gestation).
>
> (Kull et al. 2009, 169)

The function of hemoglobin is not intrinsic to its chemical or cellular structure but is related to a constitutive absence defined by the physical and historical context in which it acts; "In effect, the missing oxygen with respect to which the hemoglobin structure has evolved has become its defining characteristic" (Kull et al. 2009, 169). From this point of view, "one can understand the structure of hemoglobin as a 'representation' of both oxygen and its role in the cellular molecular processes of metabolism" (Kull et al. 2009, 169). The sign is a way of extending and strengthening life.

Because an organism must incessantly remake itself utilizing resources afforded by its environment, "it must be in dynamical correspondence with these crucial intrinsically absent features, and at the same time its constituent parts and dynamics must be reciprocally generating one another with respect to this absence" (Kull et al. 2009, 171). The function develops in relation to an end: to survive. The need for semiosis arises from this relationship because the end is absent—it is a future possibility, a possible evolution of things. Life is characterized by an emergent evolutionary history in which the semiosis becomes increasingly more differentiated and more complex. The concept of "threshold" is essential: semiotic processes are an emergent property of the interaction between chemical processes and the context. "This implies that not only organisms evolve in the course of time, but that evolution itself evolves as well" (de Mul 2021, 2).

In summary, scholars in biosemiotics claim that "an organism is a sign-interpreting process that can be described as a recursive self-referential sign production process, dependent on or influenced by some external factors likely to be present in its environment" (Kull et al. 2009, 171). Seven properties or conditions can be isolated that must be met in order for an organism to develop semiotic processes: (1) agency, (2) normativity, (3) teleology, (4) form generation,[12] (5) the differentiation of a sign vehicle from the dynamics of the reciprocal form-generating process, (6) the categorization of signs, and (7) the inheritance of relations.

The Material Conditions of Meaning

Within this theoretical framework, adaptation is a key feature because it involves two other important concepts: selection and habit. "Adaptation also involves the selective semiotic recruitment of those physicochemical aspects of the organism environment that are relevant to the persistence of that process" (Kull et al. 2009, 170). This idea refers to two important fathers of biosemiotics: Giorgio Prodi and Jacob von Uexküll. The first describes the concept of the "natural condition of selectivity," which is the theoretical presupposition of the notion of "umwelt," created by the second.

Prodi (1988) states, clearly and effectively,

> The usual perspective must be reversed. We consider our codes, and we wonder how closely nature follows them. In this way we elaborate metaphors about nature. Instead, it is necessary to

> observe nature, establish its fundamental codes, and see ours as their specialization.
>
> (Prodi 1988, 929; my translation)

For Prodi, biology is natural semiotics. What we call *life* arises from a natural condition, namely, selectivity. In the natural world there are repetitive—and non-selective—cyclical states of fact. In these states of fact, thanks to the formation of specific conditions (e.g., a certain temperature or the possibility of molecular kinetics), selective, non-repetitive phenomena can develop. Conditions of selectivity, for example, exist when certain molecules reach a level of stability and complexity at which they become capable of interacting only with a certain type of molecule and not with others. By virtue of their constitution, these molecules "choose" which molecules to join and transform with. "The condition of selectivity," writes Prodi, "arises in nature when there are molecular formations sufficiently developed in space and sufficiently stable, capable of reacting only with certain others, forming defined and reversible complexes"; further, "the chemistry of carbon is the only one that offers such conditions of stability and complexity combined with the presence of weak bonds" (Prodi 1988, 930; my translation). For Prodi, the fundamental condition of life is the carbon atom because only the carbon atom has adequate levels of stability and complexity to allow selectivity—that is, the ability to react and form a complex with other structures.

Prodi gives the example of the enzyme and the substrate—the substrate is any molecule on which an enzyme can act. The enzyme is a protein; the atoms composing it give it a particular steric configuration, meaning a specific capacity to interface with determinate atomic configurations, excluding many others (Cimatti 2018, 52). The enzyme explores its environment and interacts only with a certain substrate; it maintains a complementary relationship with it. In the enzyme environment, the only thing the enzyme can "read" and interact with is that specific substrate. Complementarity corresponds to the reaction, to the exchange of energy, that the enzyme produces only when it meets a certain substrate and "reads" it. This point must be stressed: when Prodi talks of a "reader or interpreter," "he is not thinking that the organism would be capable of reading the world without any constraint" (Cimatti 2018, 39). The organism "reads" the world by seeking complementary things—that is, objects with which it can establish stable and meaningful links. The complementary object is at the same time a sign and a referent.

As Prodi claims,

> The enzyme is selective toward its substrate because it actually reacts with this (and only with this) forming a complex. Two natural terms (enzyme and substrate) are thereby correspondent or mutually related through a reaction which is a selectivity function. There is no "active

> choice" referable to our human concept of choice: any presence that does not correspond to the substrate is left aside, does not give rise to reaction, while the substrate gives rise to reaction. The selectivity situation is revealed by (or translates into) an availability of energy (with the formation of the complex): this relationship is "targeted," that is, it is triggered by another natural presence (the substrate) and only by that one.
>
> (Prodi 1988, 931; my translation)

Therefore, "at the beginning of semiosis, there is a selective material operation, wherein a certain material configuration is 'preferred' to another. All the other forms of semiosis derive from this fundamental operation" (Cimatti 2018, 35). Selection is a condition of meaning, what Prodi calls the "natural condition of meaning." What is selected, in relation to which a reaction occurs, is significant. The substrate is significant for the enzyme because their union produces a reaction. In contact with the substrate, the reality for the enzyme stops being indifferent—a difference is produced. The enzyme can "read" the reality, recognize a difference, and identify the "thing-meaning," the substrate, as well as what produces a reaction in it and modifies it. The reader-enzyme categorizes reality, in the double sense that all enzymes react in the same way with that substrate and that the enzyme itself relates to the substrate as a whole. The enzyme categorizes the surrounding world by individuating classes of things with which it is possible to interact, setting them apart from those with which interaction is impossible—which are indifferent: "A thing is a sign when there is a natural identification system for it, that is, a reader for whom that thing is significant (term for a reaction of selectivity)" (Prodi 1988, 932). Prodi's theoretical proposal, then, "does not presuppose the existence of any intentional process. Semiosis, Prodi argues, *does not need a subject or any psychological intentionality*" (Cimatti 2018, 35; emphasis added). Semiotic processes are not an exclusive prerogative of human beings; semiosis is the result of certain biological conditions.

Another example is that of the bee and the flower. There are biological starting conditions: the bee and its needs due to its physical structure—these needs dictate the ends. These starting conditions define a natural selectivity; the bee joins the flower because the flower matches its structure and needs. The flower is significant, is important, for the bee—a difference is introduced in the environment. The flower also becomes a sign of something missing: survival, the future. The bee can "read" the flower and understand how it can fulfill that purpose.[13]

The natural semiosis model defined by Prodi directly contrasts with other semiosis models, which can be defined as "Cartesian" because they are implicitly based on the mind/brain, or nature/culture, dualism. A Cartesian model of semiosis, for example, is the one on which the information theory of Shannon and Weaver is based. This model is linear and mentalistic. It is linear because information is always a process that goes from a source to a recipient and is defined as being opposed to noise—all that is not noise is information. It is mentalistic because the ultimate source of information is always the human mind. As Shannon and Weaver (1964) write, "The word communication will be used here in a very broad sense to include all of the procedures by which one *mind* may affect another" (10; emphasis added). Although Shannon and Weaver explicitly state that their model can also be applied to automatic systems, for them, the ultimate source of information is the human mind. It is human intentionality that establishes the distinction between information and noise, the codes by which to translate signals, and so on. In other words, the ultimate foundation of semiosis is always the human being because only the human being is able to produce signs and meanings, which are not natural things. The creation of meaning involves a distance from nature, conceived as a brutal, immediate, mute fact. There can be no natural explanation of the meaning. This is also the position of phenomenology, for which it is the subject that makes sense of the world. For example, Ricoeur (1969) argues that the ontological condition of the human being is defined by a double movement of distancing from reality (i.e., the creation of meaning) and belonging (i.e., the return to reality through meaning).

Following Peirce's semiotics, Prodi claims that it is not the mind that produces semiosis but semiosis that produces the mind; the mind emerges from semiosis. Semiosis emerges from biological conditions that have nothing intentional or conscious about them: "We can . . . identify the space of signs with that of biology and interpret this as a reservoir of systems of symbols" (Prodi 1977, 48). Human intentionality and consciousness are only a small part of the infinite semiosis. For Prodi, relationship, selection, and meaning are closely connected concepts: the interaction between molecules creates a relationship, and the relationship creates a selection—that is, a distinction, a difference in the surrounding world. What is selected is significant for that organism or for that network of organisms. The conditions of these three processes are three other phenomena: matter, complexity, and evolution. Semiosis arises by emergency from an increase in molecular complexity. "Semiosis, at first, is a selective physical contact between things" (Cimatti 2018, 39), and the model is not linear: the information does not proceed from the human mind to the recipient (human or non-human). There are starting material conditions that produce a certain complexity in the matter. The increase in complexity creates differentiation and, therefore, meaning.

A clarification must be made, however. Prodi does not want to define a theory of signs but to identify the material and biological basis of semiosis. Semiosis, therefore, is not mediation, as in the work of Peirce, but a biological condition:

> Peirce does not need to postulate the intentionality and conventionality (i.e. the artificial character) [of semiosis]: however, in the way he articulates the problem of semiosis, the sign is something already given as a mediator, already part of a semiotic function, the genesis of which remains completely obscure. It is therefore necessary to go beyond: not simply to abolish intentionality, but—at the most basic stage of the process of signification—to abolish mediation itself.
>
> (Prodi 1977, 158)

According to Prodi, Peirce's reflection is still too marked by anthropomorphism. Nonetheless, it is also possible to use Peirce to go beyond Peirce:

> [T]he demarcation of the field of semiotics is a crucial point. According to de Saussure's foundation semiotics is the science of artificial and conventional signs, like language and other rule-bound systems of inter-human communication (like for example rules of politeness, traffic laws, military signs, and so on). From this point of view Peirce characterizes a generic situation, not necessarily a human one, since the process of semiosis takes place whenever a mediation between an interpreter and a thing—by means of an interpreter—obtains. But in Peirce's framework . . . the only possible domain for this semiosic process is a human one, or at least the act of interpretation is always configured as anthropomorphic and anthropocentric.
>
> (Prodi 1977, 158)

Prodi's position can be defined as a "semiosic materialism" (Cimatti 2018, 42). The sign is a "thing" that "doesn't send back towards an indefinite chain (a sign explained by a sign, explained by another sign and so on) but a chain with a finite number of interactions" (Prodi 1977, 158)—this position is therefore different from Peirce's infinite semiosis. The model of natural semiosis is that of the complementary relationship between deoxyribonucleic acid (DNA) and the amino acids that compose proteins; "In this example no mediation takes place: on the contrary, to every sequence of nucleotides corresponds a specific amino acid" (Cimatti 2018, 42). This is the natural genesis of semiosis: "The correspondence rules between DNA and proteins . . . represent the most conspicuous and general example of this historical interpretation of meaning, what I have called natural semiotics, upon which the whole of biology is founded" (Prodi 1989, 36–37). The presupposition of this approach is the rejection not only of Cartesian dualism, as mentioned, but also of the notion of nature as a simple innate datum, as passive matter opposed to the human subject. This idea paves the way toward a non-anthropocentric conception of *agency*.

Prodi (1982) opens a different path, a symmetrical ontology very similar to that of Latour, in which humans and nonhumans are placed on the same level, but one which goes beyond Latour, as it solves one of Latour's crucial problems: How do associations take place? How are networks or collectives formed? Answering these questions using semiotics inspired by human language, as does Greimas, inevitably betrays the initial inspiration of symmetric ontology. A biological conception of semiosis solves the problem at its root. Biosemiosis is that "same level" on which humans and nonhumans find themselves and connect. The same distinction between humans and nonhumans arises from biological proto-semiosis. The complementarity between A and B (e.g., the enzyme and the substrate) gives birth to a new individual who, in turn, will explore its environment, select complementary individuals, and create new individuals. In this way, ever more complex semiotic relationships develop. Prodi is always careful "to preserve the principle of continuity, and therefore he has no other choice but admitting that the complex is nothing but a quantitative complication of the simple" (Cimatti 2018, 55). According to Prodi,

> [T]he continuum between things and interpreter, between nature and culture, the noumenon and its semiotic-phenomenal correlate is the foundation of knowledge, and is expressed by saying that the reader is derived from his reading world. . . . In substance, to communicate does not mean to intervene into extra-semiotic circumstances, but rather to immerse ourselves into a world that is always-already semiotic, and that has generated us as readers.
>
> (Prodi 1977, 164–65)

What Is an Umwelt?

The concept of umwelt, created by von Uexküll, is considered one of the most important for biosemiotics. For Sebeok, von Uexküll was the "chief architect" (2001a, 70) of biosemiotics. The closest English translation of *umwelt* is *model*: "All organisms communicate by use of models (umwelts, or selfworlds, each according to its species-specific sense organs), from the simplest representations of maneuvers of approach and withdrawal to the most sophisticated cosmic theories of Newton and Einstein" (Sebeok 2001b, 21–22). Most importantly, the umwelt of a living organism comprises the circulation and receiving, insofar as it is physically allowed, of signs.

Inspired by Kant, von Uexküll's umwelt describes how the physiology of an organism's sensory apparatus shapes its active experience of the environment. An *umwelt* is a "subjective environment," in the sense that each of its components has a meaning for an organism—in Prodi's sense, that is; the relation between von Uexküll's conception of umwelt and Peirce's semiotics is actually more complex (Augustyn 2009). Von Uexküll's key point was that "neither the individual cells nor the organisms are passive pawns in the hands of external forces. They create their own umwelt and in so doing become a subjective part of Nature's grand design" (Hoffmeyer 1996, 56). The umwelt theory "tells us that it is not only genes, individuals, and species that survive, but also—and perhaps rather—patterns of interpretation" (56). This is a crucial principle for all biosemiotics: organisms create their environment; they do not just adapt to it.

The question of umwelt has been taken to the extreme by Sloterdijk, the real philosopher of design of our age (Latour 2008). Sloterdijk (2011–2016) poses an elementary but radical question: *where are we?* What is the envelope that we built to inhabit the world? Being alive means building spheres, envelopes that are necessary for protection—immunology = understanding. Thinking about umwelt is thinking about spatiality and this brings us back to Earth, in contact with the climate and the atmosphere. We only really understand technology if we understand it as an attempt to broaden and strengthen the spheres—our spaces, envelopes.

According to biosemiotics, an organism creates its own umwelt by repeatedly interacting with the world. It simultaneously observes the world and changes it through interpretative acts—abductive acts, according to Sebeok (2001a). Umwelt differs from the concept of ecological niche; while ecological niches are considered objective units of an ecosystem that can be quantified, an umwelt, in contrast, is subjective and not accessible to direct measurement. In an umwelt, functional cycles are created between organism and environment. The univocal and rigid Darwinian concept of adaptation falls away; the way an umwelt is created is semiotic. The concept of umwelt is based on Peirce's semiosis and contrasts with positivist scientific methodology, for which the goal of the scientific endeavor is to discover the various aspects of the objective real world that exists independently of each observer (von Uexküll 1909, 2010; for a critical survey of von Uexküll's concept of umwelt, see Feiten 2020).

However, the theory of umwelt also has some limitations. Brentari (2016) identifies four significant problems with it: (1) what he calls von Uexküll's "Kantian problem," (2) von Uexküll's idea of perfect harmony in the relation between organism and environment, (3) the theory's relation to animal psychology, and (4) von Uexküll's political use of the umwelt theory (Brentari 2016, 9). I will not here analyze these four aspects or give an evaluation of each one, despite their importance, as this is not the aim of the present Chapter. However, the fundamental point that emerges from Brentari's interpretation is that Kantian assumptions prevent von Uexküll from fully developing the concept of umwelt. For example, von Uexküll's idea of perfect, pre-established harmony in the relation between organism and environment makes it impossible to understand the misunderstandings or short circuits that arise. As Tonnessen (2009) underscores, "umwelt theory suffers from its reliance on von Uexküll's false premise that the environment (including its mixture of species) is generally stable" (48). For this reason, Tonnessen introduces the idea of "umwelt transition," which is particularly interesting because it represents a concrete update of von Uexküll's umwelt theory:

> An umwelt transition can tentatively be defined as a lasting, systematic change, within the life cycle of a being, considered from an ontogenetic (individual), phylogenetic (population-, species-) or cultural perspective, from one typical appearance of its umwelt to another. An umwelt transition, in other words, can be regular, irregular or a singular, extraordinary event. In the last case we are entitled to talk about historical events. In a similar vein, transitional umwelten can be taken to refer to umwelten undergoing an umwelt transition or umwelten in so far as they typically go through a certain kind of umwelt transitions.
>
> (Tonnessen 2009, 49)

Tonnessen (2009) distinguishes three levels of an umwelt: the core, mediated, and conceptual levels. The core level is that of direct encounters between a subject and the objects that make up its umwelt. The mediated level, in contrast, is that of indirect encounters, meaning that it is mediated by other objects, such as artifacts, tools, or representations (memory, imagination, anticipations, etc.). The conceptual level, which is not present in all living beings, instead concerns symbolism—the relationship with reality mediated by language and cultural habits. In an even more analytical way, Tonnessen (2009) further distinguishes six levels of the umwelt:

Core Umwelt

1) Automated acts of perception
2) Automated mental acts

Mediated Umwelt

3) Willful acts of perception
4) Willful mental acts

Conceptual Umwelt

5) Habitual acts of perception
6) Habitual mental acts

There is therefore no single umwelt but many. An umwelt is constantly evolving. Semiotic processes on each level are based on codes. According to Tonnessen (2009, 45), a *code* is a set of rules that establishes a correspondence (or a mapping) between two independent worlds; it is a set of functions. Examples of this type are, in the biological world, the genetic code and the signal transduction codes (i.e., the process by which cells transform the signals from the environment, called *first messengers*, into internal signals, called *second messengers*). According to Tonnessen (2009), the umwelt core is code-based, and this code is fixed and immutable. It is a set of neural codes or rules embedded in the nervous system. The mediated and conceptual levels are also code-based, but their codes are ecological codes. Tonnessen draws on the idea of "ecological codes" from Farina (2014). *Ecological codes* can be defined as mechanisms that establish an arbitrary set of connections between two or more components (i.e., organisms and/or their aggregations) of a complex system: "The ecological codes are the tools that organisms use in everyday life to relate themselves with the external world. Ecological codes are visual, acoustic, tactile, chemical and cultural and exist at every scale of the living organization" (Brentari 2016, 15).

Criticizing Sebeok's approach because it is too tied to Peirce's semiotic model,[14] Barbieri has introduced a new form of biosemiotics: the code biology. Barbieri (2015) claims that the existence of many organic codes in nature is an experimental fact that has extraordinary implications. The history of evolution depends on the creation and preservation of new codes of living beings, from cells to bacteria, from animals to humans. This is a process that Barbieri (2012) calls *codepoiesis*. The idea of code implies that of meaning: "The ability of the cell to conserve its own codes accounts for the fact that the organic codes are the sole entities that have been perpetuated in evolution. They are the great invariants of life, the sole entities that have been conserved while everything else has changed" (Barbieri 2015, XV).

Starting from the different approaches and biosemiotic theories that I have outlined, we can draw a unitary diagram of the semiotic constitution of the living organism (see Figure 1.2). As explained, there are many different approaches to biosemiotics; nevertheless, it is possible to find a common theoretical basis.

I would like to add a final consideration. There is also another point of view from which we can re-conceptualize the notion of umwelt: that of memory. The umwelt, individual or public, is the space of memory. As Stiegler writes:

> The *interior milieu* is social memory, the shared past, that which is called "culture." It is a nongenetic memory, which is exterior to the living organism qua individual, supported by the nonzoological collective organization of objects, but which functions and

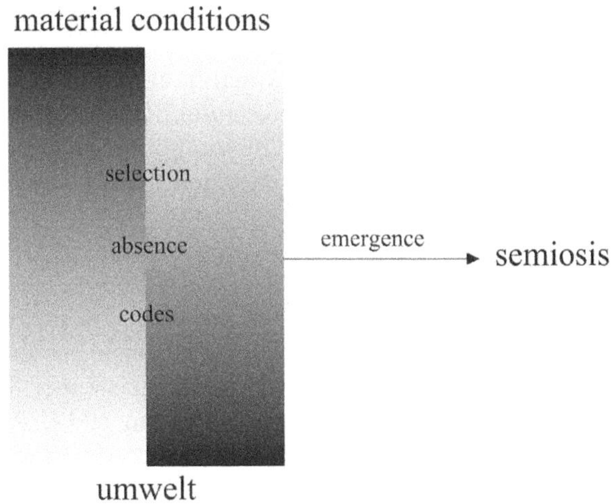

Figure 1.2 The main concepts of biosemiotics to explain the material dynamics of the constitution of meaning.

> evolves as a quasi-biological milieu. . . . The *exterior milieu* is the natural, inert milieu, but also the one carrying the objects and the ideas of different human groups. As inert milieu, it supplies merely consumable matter, and the technical envelope of a perfectly closed group will be the one allowing for the optimization of the "interior" milieu's aptitudes.
>
> (Stiegler 1998, 57; I emphasize).

Therefore, the umwelt is a set of semiotic structures that are at the same time traces of the past and anticipations of the future and that are externalized in matter.

Biosemiotics and Language

Terrence Deacon has tried to unify biosemiotics, cognitive science, neuroscience, and research on the origin of language. My claim in this section is that the main concepts of Freud's psychoanalysis can be reformulated in Deacon's terms.

The Symbolic Species (Deacon 1997) presents a theory of the origin of language. In this book, language is seen as exclusive to humans—humans being the symbolic species—and language is different from communication, which is used by all living forms. Language is exclusive to humans because a mind capable of symbolic representation is developed only in humans. Brain and language evolved together; biological evolution and cultural evolution were synergistic. Deacon clearly criticizes Darwinism:

evolution cannot be the effect of a rigid and linear process of natural selection that goes from the least perfect to the most perfect—it is instead a much more complex, articulated, and reticular process. Peirce—whose pragmatism is incompatible with Darwinian realism—is Deacon's main source of inspiration; evolution is a biosemiotic and creative process.[15] For the purpose of this chapter, two major points in Deacon's views should be discussed: the logic of emergence and the efficacy of absence.

The Logic of Emergence

Deacon (1997) takes an anti-objectivist approach: "The correspondence between words and objects is a secondary relationship, subordinate to a web of associative relationships of a quite different sort, which even allows us reference to impossible things" (Deacon 1997, 70). He then develops Peirce's theory of signs in a new direction. Peirce identifies three types of signs: icons, indices, and symbols. In the icon, there is a relationship of similarity between sign and object. The index expresses a relationship of space—time contiguity between sign and object. In the symbol, however, the relationship between sign and object is of a conventional nature; the sign implies culture and social habits (more on this in Chapter 4). Signs are always the result of an interpretation, that is, of an inferential process (i.e., deduction, induction, or abduction). Anything can be an icon, symbol, or index according to the interpretation given to it. However, the relationships between icons, indices, and symbols are different; they do not depend only on the interpretation given to them.

To explain this point, Deacon introduces the idea of levels of interpretation in a hierarchical way. More complex semiotic forms are built on more elementary semiotic forms; more complex reference modes rely on simpler reference modes. He then individuates three levels: iconic, indexical, and symbolic. "Reference itself is hierarchic in structure; more complex forms of reference are built up from simpler forms" (Deacon 1997, 73). Therefore,

> though I may fail to grasp the symbolic reference of a sign, I might still be able to interpret it as an index (i.e., as correlated with something else), and if I also fail to recognize any indexical correspondences, I may still be able to interpret it as an icon (i.e., recognize its resemblance to something else).
>
> (Deacon 1997, 74)

Symbolic processes are grounded in simpler, more iconic, and indexical semiotic processes. The index reference depends on the iconic one, while the symbolic reference depends on the index one:

> What we really mean is that the competence to interpret something symbolically depends upon already having the competence to interpret many other subordinate relationships indexically, and so forth. It is one kind of competence that grows out of and depends upon a very different kind of competence. What constitutes competence in this sense is the ability to produce an interpretive response that provides the necessary infrastructure of more basic iconic and/or indexical interpretations. To explain the basis of symbolic communication, then, we must describe what constitutes a symbolic interpretant, but to do this we need first to explain the production of iconic and indexical interpretants and then to explain how these are each recoded in turn to produce the higher-order forms.
>
> (Deacon 1997, 74)

According to Deacon, symbolic relationships are composed of indexical relationships between sets of indices, and indexical relationships are composed of iconic relationships between sets of icons. This is a hierarchical model based on a "logic of emergence" (Deacon 2003). Icons and indices are semiotic relationships based on material and biological dynamics that all living beings can produce and understand. The increase in complexity in matter allows the emergence of semiotic relationships and, therefore, of icons and indices. Semiotic relations are new forms of organization in matter; they present new qualities compared to the previous levels, even if they cannot exist without the latter. What we call *meaning*—linguistic denotation and connotation—belongs only to the symbolic and cultural level, not to the indexical or iconic ones. It is a new organizational level that only the human mind has been able to create. In language, signs are defined in relation to each other, abstracting from the physical environment. To learn the symbols, we must start from icons and indices, semiotic relationships that we share with animals and plants; through icons and indices, we learn the relationships between signs and objects, but once learned, these associations become only clues to define more crucial relationships.

Two aspects should be emphasized in Deacon's position. First, according to Deacon's vision, as Kohn (2013) also points out, the symbolic is in perfect continuity with simpler semiotic levels such as icons and indices, which are shared by humans and nonhumans. Second, "understanding something, however provisional that understanding may be, involves an icon"; it involves "a thought that is like its object" (Kohn 2013, 51). For this reason, "all semiosis ultimately relies on the transformation of more complex signs into icons" (Kohn 2013, 51).

Let us try to make the relationship between indices and symbols clearer (see Deacon 1997, 85–100). First, within a set of indices, each index is learned individually. Second, systematic relationships between index occurrences are recognized and learned as additional, supplementary indices. Third, a change in mnemonic strategy occurs such that the relationships between occurrences of the indices are assumed as tools to recognize objects and the relationships between objects indirectly. Memory associates the occurrence of indices with the occurrence of objects and uses these associations to refer to objects indirectly. Symbolic relationships are nothing more than abstract indexical relationships. In other words, in the passage from the indexical to the symbolic level, a radical change occurs in our way of using signs; we no longer use signs to indicate every single object but to indicate relationships between objects and between signs themselves. Symbols do not refer directly to the world but indirectly, through other symbols in combinatorial relationships. Many animals have limited symbolic abilities. "Symbolic reference emerges from a ground of non-symbolic referential processes only because the indexical relationships between symbols are organized so as to form a logically closed group of mappings from symbol to symbol" (Deacon 1997, 99). From indices to symbols, the brain moves from an associative strategy to a systematic strategy. This system of relationships between symbols "determines a definite and distinctive topology that all operations involving those symbols must respect in order to retain referential power" (Deacon 1997, 99). The development of symbolic relationships brings out new properties and combinatorial relations.

Human symbolic capacity arises through an emergent dynamic from the intertwining of numerous indexical and iconic relationships. It cannot exist without this network of indices and icons, even if it has entirely new characteristics, which do not appear in the other levels of interpretation. Deacon uses the concept of emergent dynamics to describe the behavior of complex systems capable of self-organization: "Complex dynamical ensembles can spontaneously assume ordered patterns of behavior that are not prefigured in the properties of their component elements or in their interaction patterns" (Deacon 2003, 274). Moreover, "unprecedented global forms can develop along parallel lines to reach similar patterns of behavior despite arising from components of radically different constitution, interacting according to quite diverse physical principles" (Deacon 2003, 275). What emerges is not a thing but a property, a behavior, or a habit. Deacon presents several examples: "Whatever we mean by 'emergence,' there can be no doubt that mental phenomena are emergent from the subordinate neuro-chemical interactions occurring in a brain in a more complex way than liquid phenomena are emergent from water molecule interactions" (Deacon 2003, 279).

The Efficacy of Absence

According to Deacon (2011), present-day science is incomplete, as it does not include human feeling, attitude, hope, value, and purpose—for which he coined the term *ententions*. His revolutionary proposal is to include the concept of absence in science, just as the inclusion of zero as a placeholder or symbol in the Middle Ages led to the Arabic numbering system that we find so useful today. Absence is full of potential, as is the void within a glass container. The central question is how can something not physically there (i.e., something ententional) be the cause of anything? Deacon (2011) develops the idea of "efficacy of absence," so that ententions become an integral part of science and information theory.

Absence is an integral part of Deacon's logic of emergence, which explains how the first cell came from matter by natural processes. In the conventional understanding of emergence, primitive cells emerged with novel properties that are greater than the sum of their interacting parts. Deacon suggests the opposite: novel properties can be less than the sum of their parts. However, for him, "less is more"; absence is a constraint that limits each part's infinite number of possibilities to the function that contributes to the whole. Deacon distinguishes three stages leading to the emergence of the first living cell from matter:

1 *Thermodynamics*: Atoms and molecules of water, methane, ammonia, carbon dioxide, and so on move randomly from thermal fluctuations in a primordial soup.
2 *Morphodynamics*: This involves the emergence of a self-organizing form, or "order for free," and the absence of dynamical variety. For example, diamond crystals found in the earth have carbon atoms with an orderly cubic structure. At high temperatures and pressures, diamonds emerge from the self-organization of clusters of carbon atoms in the earth. Artificial diamonds are made using the same high temperatures and pressures. Morphodynamics also includes stable processes, like the flow of a river; the overall shape, or the river's form, remains the same even though each water molecule is continually flowing downstream. In autocatalytic chemical processes, the output products feed stably back into the input.
3 *Teleodynamics*. Living cells emerge under the right conditions from amino acids, proteins, and autocatalytic processes in the primordial soup. The vital purpose (i.e., telos) of a cell is to eat and to avoid absence (i.e., to avoid being eaten), as well as to reproduce. The behavior and development of cells are constrained by absence. Each part is constrained to a function that serves the whole. To survive, a cell must move away from areas where food is absent to those where it is present. This is the semiotic level. Icons and indices appear here.

Deacon develops an emergent theory of energy and work. He applies the emergent steps of thermodynamics, morphodynamics, and teleodynamics to the playing of a flute: thermodynamics represents the energy that the player expends in blowing the flue; morphodynamics is the vibrational patterns of the standing waves of sound within the instrument; while teleodynamics is the meaning and purpose for which the flute is played. The mind emerges from the constraints determined by the absence in material conditions. This allows me to underscore a point I have mentioned several times earlier: semiosis is the way living beings manage absence and the constraints posed by the absence (Figure 1.3).

Now, let us go back to Freud. My question is, can we interpret repression as the impossibility of translating a complex of icons and indices into symbols? In his essays on *Metapsychology*, Freud distinguishes original repression from actual repression. The first affects the representation and the connected drive in such a way that both continue to exist in the unconscious but not in the conscious. For this, Freud states, the original repression produces a fixation. The actual repression is instead the result of two forces: the first is that of the ego censorship, while the second concerns the attraction exerted by the original repression on everything that can be associated with it. Nothing prevents us, respecting Freud's text, from thinking that, in both cases, the representation connected to the drive is an icon generated by the organic stimulus, a very simple and biological form of a sign. Freud clearly states that the representational modality of the unconscious is not linguistic but visual. The repressed icon generates a chain of signs (i.e., icons or indices) along which the drive energy is distributed. The concatenation of the levels identified by Deacon is blocked by the repression; it is not possible to organize and build an appropriate symbolic level. The removed icon remains present

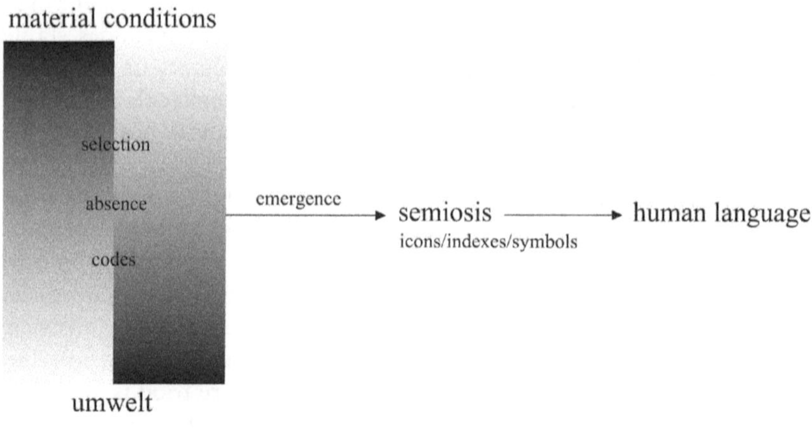

Figure 1.3 The main concepts of biosemiotics with Deacon's integrations.

but cannot be verbalized. The unconscious is semiotic in the sense that it is made up of chains of icons and indices that do not translate into symbols, and—when the ego's control fails—they show themselves bypassing censorship and deforming the symbolic. The analyst's job is to analyze the symbolic chain and trace it back to the lost icon and the underlying work of the drives. This is perfectly in line with what Freud states in *The Ego and the Id*:

> The real difference between a Ucs. and a Pcs. idea (thought) consists in this: that the former is carried out on some material which remains unknown, whereas the latter (the Pcs.) is in addition brought into connection with word-presentations.
>
> (Freud 1960, 12)

Language represents, for Freud, a phylogenetically recent development of the human psychic apparatus. Thanks to the language, the human psychic apparatus further develops the typical characteristics of secondary processes (on psychoanalysis and language, see Forrester 1984; Salvador 2020).

What happens when a representation is translated from the unconscious to the conscious system? For Freud (1957, 155), the unconscious representation must be connected to preconscious verbal elements to be translated into consciousness. *Conscious* representation is the representation of the thing plus the representation of the word; *unconscious* representation, however, is only the representation of the thing (Freud 1960, 15–18). Freud specifies that the conjunction with verbal elements does not exactly coincide with the passage to consciousness but only constitutes its condition of possibility. However, the fundamental function of language is to allow the awareness and control of internal thought processes. Through thought, the ego can influence psychic dynamics. Repression concerns the possibility that an unconscious representation may or may not translate into preconscious verbal representations. The repressed is the representation of the thing without the representation of the word. Because language is an over-investment, repression corresponds to a counter-investment. Following our biosemiotic interpretation, the repressed is a set of icons and indices that cannot be translated into symbols.

Biosemiotic interpretation gives us a crucial advantage: it highlights Freud's central intuition, namely, that language is basically a series of acoustic (and partly visual) signs whose purpose is not communication but rather the management and organization of psychic processes. Language always remains at the service of thought. What is thought?

> It is probable that thinking was originally unconscious, in so far as it went beyond mere ideational presentations and was directed to the relations between impressions of objects, and that it did not acquire

> further qualities, perceptible to consciousness, until it became connected with verbal residues.
>
> (Freud 1958, 221)

Thought is essentially a process of postponement of instinctual satisfaction. This process has a practical and biological function; in fact, the postponement of instinctual satisfaction allows the mind not only to plan for satisfaction in the future, and therefore to make it simpler, but also to realize it in increasingly complex situations: "Thinking was endowed with characteristics which made it possible for the mental apparatus to tolerate an increased tension of stimulus while the process of discharge was postponed" (Freud 1958, 221). Through the postponement of the satisfaction of the drive, the process of the maturation of the individual takes place; it is the transformation of the pleasure principle into the reality principle:

> The aim and end of all processes of thought are the establishment of a state of identity. . . . Cognitive or judging thought seeks for an identity with a somatic cathexis; reproductive thought seeks for an identity with a psychical cathexis (an experience of the subject's own). Judging thought operates in advance of reproductive thought, since the former furnishes the latter with ready-made facilitations to assist further associative travelling. If at the conclusion of the act of thought the indication of reality also reaches perception, then a judgement of reality, a belief, is achieved and the aim of the whole activity is attained.
>
> (Freud 1953, 394–395)

Freud underlines three aspects of thought and language: (1) their associative nature, (2) their practical purpose, and (3) their energetic dimension, which is their connection with unconscious drive dynamics. In Freud, there are—as in Peirce—two distinct notions of meaning: (1) meaning as a representation of the thing, the immediate object of the sign; and (2) meaning as drive—that is, the drive toward the dynamic object, following Peirce's terminology. Therefore, Freud's conception of thought and language is focused on the biological dimension. Thought and language are the first survival strategies or ways of achieving the satisfaction of stimuli. This does not require reducing the mind to a stimulus—response model; instead, it requires maintaining the conceptual capacities of the mind and at the same time rooting them in the biological substratum, in the life of the drives. "For this reason, the study of linguistic phenomena such as slips allows us to reconstruct the underlying unconscious dynamics, in the absence of which these same phenomena would remain without real explanation and meaning" (Salvador 2020, 209; my translation). The sign is alive; it is rooted in biology. For this reason, the living organism can be modified by the sign.

Interpreting repression from a biosemiotic point of view does not mean wanting to propose abstract pseudophilosophical formulas that betray the complexity of Freud's thought. On the contrary, my intent is to show the possibility of a different perspective in Freud's interpretation, far from structuralism and its emulators. My approach, based on Deacon's biosemiotics, intends to show that language is not a set of abstract forms that would make up an autonomous and ideal world but rather something strongly rooted in animal and human biology and its evolution.

There is also another crucial point that, in my opinion, should be highlighted. In the text on aphasias of 1881, Freud criticizes the localization model, which is the theory that explained aphasias by likening them to physical lesions on parts of the brain. Underlining the importance of clinical observation, Freud states that the localization scheme is too simple and schematic; it takes into consideration the activity of language as repetition but not as spontaneous creation (Freud 1953, 60). The description of aphasias must be based on (1) clinical analysis of the lesions, (2) brain physiology, and (3) a general psychological theory of the functioning of language. Freud, therefore, denies that the psychic and physiological are identical, even if he does not eliminate any relationship between them. He states that physiological processes are not in a causal relationship with psychic processes, but he does not contend that there is no relationship at all. This does not compromise our biosemiotic interpretation; Deacon's emergency logic is not a causal logic. As Freud writes,

> From the psychological point of view the "word" is the functional unit of speech; it is a complex concept constituted of auditory, visual, and kinesthetic elements. We owe the knowledge of this structure to pathology which demonstrates that organic lesions affecting the speech apparatus result in a disintegration of speech corresponding to such a constitution. We have learned to regard the loss of any one of these elements as the most important pointer to the localization of the damage. Four constituents of the word concept are usually listed: the "sound image" or "sound impression," the "visual letter image," the "glossa-kinaesthetic and the cheirokinaesthetic images or impressions." . . . The word, then, is a complicated concept built up from various impressions, i.e., it corresponds to an intricate process of associations entered into by elements of visual, acoustic and kinaesthetic origins. However, the word acquires its significance through its association with the "idea (concept) of the object," at least if we restrict our considerations to nouns.
>
> (Freud 1953, 73)

Language is constituted by a network of representations connected to organic processes—the stimuli. Freud distinguishes two types of representation: the representation of the word and the representation of the object. These two representations are, in turn, composed of a network of visual, acoustic, and tactile representations. The network of the words is closed, while that of the objects is open. Sound plays a central role in the representation of the word, while visuals play a central role in the representation of the object. In Chapter VI of his essay on aphasia, Freud states that when we learn to speak, we always start by hearing a sound that we try to replicate. We associate that sound with that word. The replicated sound is an icon of the original one. Other iconic relationships are then grafted onto this icon between groups of acoustic elements and between groups of acoustic and visual elements. Onto this iconic network are then grafted ever more complex indexical and symbolic processes because they are aimed at managing ever more complex drive dynamics.

I will next summarize this interpretation of Freud through biosemiotics. My general thesis is that psychoanalysis is a natural science based on (1) a method of experimental verification with its own specific logic and (2) a biosemiotic approach to the study of the human mind. The biosemiotic interpretation does not betray Freud's thought; in fact, it strengthens it, giving it a strong empirical basis. Through the interpretation of the symbolic associative chains provided by the patient, the analyst tries to go back to the indexical and iconic structures that support them and, therefore, to the unconscious drive dynamics, which are the true meaning of those structures. In doing this, the analyst uses the theoretical constructions that they must verify through the MES. The biosemiotic assumption is essential for Freud because metapsychology and the analyst's own action would not be possible without it—the therapy would not make sense; it would be useless chatter. From this point of view, the MES can be seen as a technique for analyzing and transforming semiosis, protecting its unity, and revealing its origin and organic conditions.

The connection between psychoanalysis and contemporary research in biosemiotics enriches Freudian investigation because it opens the doors to areas of investigation that were previously foreign to it. The unconscious is no longer only a set of private human representations and impulses closed in the skull of the human subject but something much broader and that coincides with Peirce's infinite semiosis. Psychoanalysis thus becomes a pathway to privileged knowledge of the profound link between the human being and its environment, understood as the semiosphere and biosphere, or semiosphere—biosphere (Hoffmeyer 1996). The human unconscious is nothing more than a gateway to a much larger world, in which organic and semiotic, life and signs, are connected. If, as I claim, the unconscious is the semiosis that cannot be translated into human symbolic language, then there is also an unconscious in plants, animals, bacteria, cells, and forests—to paraphrase Kohn (2013). Further, the semiosis that can be translated into human symbolic language represents only a small part of the semiosphere—biosphere. From this point of view, psychoneurosis must be interpreted as a laceration of the semiosphere—biosphere and therapy as a way to mend this laceration.

In the next section, I intend to test this biosemiotic interpretation of Freud by reinterpreting the dream "Irma's Injection."

Freud's Irma Dream

"Irma's Injection" is one of the most famous dreams in the history of psychoanalysis. There is a vast body of literature exploring and analyzing it (Kuper and Stone 1982; Sprengnether 2003; Langs 1984). In this section, I do not pretend to add anything original to this literature. I simply aim to demonstrate how Freud's work in dream analysis consists of regressing from the symbolic level to the iconic level, therefore retracing Deacon's (1997) scheme backward. The identification of the fundamental icons is the first step in the identification of drive dynamics. We already know that the meaning of the dream is the fulfillment of several of Freud's desires: the desire for revenge against Irma, M., and Otto for having accused him of not doing his job well; the desire to be acquitted of the accusations; and, therefore, the desire of the recognition of his value. In other words, the central themes are guilt and accusation.

Given the importance of the dream under consideration, I believe that the best way to begin this section is to quote Freud's full description of it:

> A large hall—numerous guests, whom we were receiving—Among them was Irma. I at once took her on one side, as though to answer her letter and to reproach her for not having accepted my "solution" yet. I said to her: "If you still get pains, it's really only your fault." She replied: "If you only knew what pains I've got now in my throat and stomach and abdomen—it's choking me"—I was alarmed and looked at her. She looked pale and puffy. I thought to myself that after all I must be missing some organic trouble. I took her to the window and looked down her throat, and she showed signs of recalcitrance, like women with artificial dentures. I thought to myself that there was really no need for her to do that.—She then opened her mouth properly and on the right I found a big white patch; at another place I saw extensive whitish grey scabs upon some remarkable curly structures which were evidently modelled on the turbinal bones of the nose.—I at once called in Dr. M., and he repeated the examination and confirmed it. . . . Dr. M. looked quite different from usual; he was very pale, he walked with a limp and his chin was clean-shaven. . . . My friend Otto was now standing beside her as well, and my friend Leopold was percussing her through her bodice and saying: "She has a dull area low down on the left." He also indicated that a portion of the skin on the left shoulder was infiltrated. (I noticed this, just as he did, in spite of her dress.) . . . M. said: "There's

> no doubt it's an infection, but no matter; dysentery will supervene and the toxin will be eliminated." . . . We were directly aware, too, of the origin of her infection. Not long before, when she was feeling unwell, my friend Otto had given her an injection of a preparation of propyl, propyls . . . propionic acid . . . trimethylamin (and I saw before me the formula for this printed in heavy type). . . . Injections of that sort ought not to be made so thoughtlessly. . . . And probably the syringe had not been clean.
>
> (Freud 1955, 131–132)

Freud's description is the symbolic level of analysis, the story of the dream; the dream is translated into linguistic terms. Now, I identify four great icons that emerge in Freud's interpretation of the dream:

1 First icon: Irma/housekeeper/friend of Irma/wife of Freud
2 Second icon: M./Freud's brother
3 Third icon: liquor/syringe
4 Fourth icon: turbinal bones of the nose/female sexual organ

Each of these icons, which constitute the "submerged nuclei" of the dream, corresponds to a work of substitution. Furthermore, each of these icons corresponds to a shift. For Freud, the shift concerns the psychic intensities of the individual elements; in a dream, the truly important elements are the marginal ones. We can identify four shifts in psychic intensities in the Irma dream and in the interpretation that Freud gives of it:

1 First shift: blame for Irma's illness → attack on Irma → Irma's friend's illness
2 Second shift: blame for Irma's illness → revenge against M. and Freud's brother
3 Third shift: blame for Irma's illness → revenge against Otto and M.
4 Fourth shift: revenge against Otto and M. → sexuality

A network of very complex associations develops around the icons and shifts in the dream narrative and in Freud's interpretation, which we can summarize as follows:

1 Reception/window/mouth/throat/dentures/Irma's white spot/turbinal bones of the nose/Freud's eldest daughter/Freud's concern for her health/cocaine/patient's illness/friend's death from cocaine abuse
2 Pallor/beard/sulfonal/death of the patient/name of Freud's eldest daughter
3 Otto/Leopold/sick children/skin/rheumatism/infection/dysentery/poison/diphtheria/mockery of colleagues

4 Injection/propyl/propionic acid/trimethylamine/printed formula/ syringe/liqueur/Otto/Fliess

Each element in these networks is an interpretant, in Peirce's sense—that is, a new sign created by a previous sign in an infinite process that stops only when a habit is established or transformed. The sign–object relationship produces new signs (i.e., interpretants), which serve to define the relationship itself or to mediate between the sign and object. However, there are not only iconic relationships, substitutions, and shifts in the Irma dream but also indexical relationships. As mentioned, the index in Peirce mainly concerns space and spatial relationships. I identify four main spatial relationships in the dream:

1 Proximity between Freud and Irma
2 Moving Freud and Irma toward the window
3 Moving into Irma's mouth
4 Proximity between Irma, Leopold, and Otto

These four indexical/spatial relationships constitute a network that overlaps the four iconic nuclei. The story of the dream grafts the language, the narrative, and the logical sequentiality on this network of icons, shifts, and indices. The analysis breaks down the associative chains and individuates icons, shifts, and indices that regulate and orient that chain. This is the interpretative work of Freud.

The patient suffers from resistances.[16] Their symptoms are substitutes for what they have repressed. The analyst tries to reconstruct that which has been repressed starting from the traces of the repressed that they find in the material made available by the patient—icons and indices. The first movement of analysis is interpretation, which is a procedure based on precise rules:

1 First rule: The interpretation must apply only to the associative material provided by the patient during the session. The analyst must never take the place of the patient or introduce their own associative chains.
2 Second rule: The interpretation must apply not to the contents of the representations but to the associative links; the repression does not concern the contents, in fact, but the links between the representations. The interpretation must restore the associative links where they are interrupted.
3 Third rule: The repressed must be current and active. This means that the associative links studied must be (a) made by the patient in the present, current session, following the first rule, and (b) active and invested, with their actualization taking place through the

transference. How do we recognize which associative links between representations "hide" the repressed and which do not?

4 Fourth rule: The associative links that "hide" the repressed are those that are incomprehensible—that is, those that require a sense, about which the patient resistance is more evident.
5 Fifth rule: The basic conceptual network of any interpretation is Freudian metapsychology.

If the analyst does not respect these rules, their interpretation is invalid, and—as often happens—it is nothing more than a projective interpretation, meaning an analyst's projection of their own contents onto the patient. Interpretation is linked to the second movement of analysis, namely, construction. Therefore, the interpretation identifies icons, indices, and shifts, which the analyst assumes are signs of the patient's drive dynamics (i.e., the resistances). Starting from this work of decomposition, the analyst formulates their etiological hypothesis, which they later test through the MES. The construction then proceeds from icons and indices to the symbolic levels, up the communication to the patient. My main claim is that the analyst's goal is to restore the continuity between the semiotic and the biological domains where this contact has been interrupted. Psychoneurosis is a laceration of the semiosphere—biosphere, and analysis is a way to mend this laceration.

Accordingly, I schematize Freudian analysis in the following terms (see Figure 1.4), completing what has been said in the previous sections and in the Introduction:

Freud's text is very clear, and in this chapter, I have only skimmed the surface of the issues that can be addressed by applying biosemiotics to Freudian psychoanalysis. In fact, there are many different types of indices and icons. There are no pure icons, indices, or symbols; there is always a mixture of these three interpretative genres. What I want to emphasize is Freud's biosemiotic approach: the work of interpretation and construction intends to restore the continuity between the semiotic and organic—the semiotic speaks to us of the organic, and vice versa. This work can be explained, or clarified, following Deacon's scheme: semiotics is an emergent property of the organic, even if it is not reducible to it.

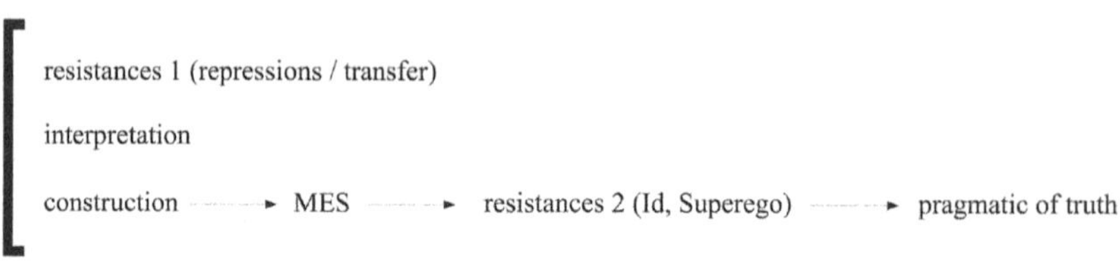

Figure 1.4 The set of procedures comprised by the Freudian method.

Conclusions

Let us try to draw some provisional conclusions from what has been said. I summarize these results in five points:

1 The mind is a hybrid of the semiotic and organic (i.e., the biosemiotic assumption).
2 If the mind is a hybrid, then the mind is a collective, an association of several semiotic and/or organic elements (i.e., the ego, the id, the superego, organic stimuli, investments, psychoneuroses, etc.).
3 Psychoanalysis investigates these types of hybrids—this is its object.
4 Psychoanalysis seeks to obtain objective knowledge of these hybrids through interpretations and constructions and by subjecting these constructions to experimental verification.
5 Starting from this objective knowledge, psychoanalysis also tries to intervene in these hybrids by modifying them, mending their lacerations.

In psychoanalysis, what we call the *mind* is a hybrid that is both semiotic and organic. The analyst's goal is not to reduce the semiotic to the organic nor the organic to the semiotic but to show the dynamic interactions between the two—that is, to show how the semiotic and the organic would make no sense outside of that relationship. Freud knows well, as biosemiotics claims, that the organic alone does not exist, because there is only the organic—semiotic. The drive is a network of semiotic and organic elements in continuous negotiation—continuously exchanging investments and counter-investments.

If the aforementioned five points are true, then biosemiotics gives us important conceptual tools to improve Freud's analysis. If the mind is a semiotic and organic hybrid, then the mind no longer concerns not only the relationship between the human and the body, but also the umwelt, in all the dimensions indicated earlier; it therefore also concerns the relationship between humans and nonhumans. I claim that human drives and organic stimuli are representations—sets of signs—of the umwelt. The drives are signs of the organic stimuli, but the organic stimuli are at the same time signs of the umwelt—the signs of the umwelt in the human body. Biosemiotics invites us to extend Freud's thinking in this direction: the mind is a collective of culture and nature, a network of human and nonhuman actants.

As we will see in the next chapter, Latour's ANT helps us develop this idea: the distinction between semiotic and organic is the result of a typical modern process of splitting—of which Freud is partly aware, although he remains deeply Kantian. The semiotic and the organic as independent realities do not exist. The semiotic and the organic are collectives of actants in perennial negotiation. We must, however, pose a few questions, which will be the focus of the next chapter:

1 Does not interpreting Freud through Latour result in the loss of the scientific nature of the Freudian method?
2 Does psychoanalysis as a science of nature fall under the Latourian critique of epistemology?
3 What kind of collective is the mind?
4 By assuming Latour's symmetrical ontology, are we not betraying Freud, who gives the human subject an exceptional place?
5 How does the mind relate to other collectives?

Notes

1 It is possible to compare two theoretical problems: that of the placebo effect in medicine and psychology and that of suggestion in psychoanalysis, as posited by Baldini (1998) and Cagna (2019). This comparison is very important when clarifying the problem of objectivity in psychoanalysis—that is, the experimental control of theoretical hypotheses. As Cagna points out (2019, 133–137), the problem of the placebo in medicine is analogous to the problem of the objectivity of the phenomenon observed in quantum physics. The position of the observer (i.e., the doctor who administers the placebo) is not neutral and influences the observed (i.e., the patient). The psychosocial component activates neurobiological mechanisms that produce a dynamic disturbance within the system (i.e., the patient). Cagna rightly emphasizes the difficulty of completely eliminating the placebo effect in medicine. It is even more difficult to identify placebo effects in psychology; it is not possible to establish whether the results of a psychotherapy are nothing more than placebo effects or whether they are the consequence of a technique, a method, or a theory. Cagna (2019) connects this problem to the historical debate about suggestion, a long-debated issue from the end of the nineteenth century to the beginning of the twentieth century, when hypnosis was experimented with as a method of treatment in psychiatry and neurology. Freud was very familiar with this debate due to his knowledge of Charcot's work.

2 *Psychoneurosis*—a term now abandoned by diagnostic manuals—is a specific phenomenon and is the object of psychoanalysis. It should not be confused with other forms of neuroses (such as organic or traumatic ones) or disorders (see Laplanche and Pontalis 1988, 266).

3 By *intelligence,* I mean, following Freud, the set of activities involved in the rationalization of phenomena.

4 "A placebo is a pharmaceutical form that does not contain biologically active substances but which can produce positive therapeutic effects based on the expectations of those who take it. It is the fact of administering the placebo that arouses the expectation; therefore, it is clear that in this case, too, as in quantum physics, the observer's position is not neutral with respect to the observed object. Of course, this is a serious problem when we have to evaluate the efficacy of a certain active ingredient in pharmacology because it is not possible to understand whether the patient's improvement is due to the active ingredient or the placebo" (Baldini 2020, 13; my translation). See also Benedetti (2008, 2015). As noted in the Introduction, the suggestion has two disadvantages: (1) its benefits tend to disappear over time and (2) it is valid only for a minority of subjects. "We therefore understand very well the importance, for a psychology that wants to call itself scientific, of finding a way to distinguish the effects due to a certain therapeutic intervention from those due to

suggestion" (Baldini 2020, 14; my translation). As Baldini (2020) points out, it is important to underline that the methods applied in medicine to eliminate the placebo effect cannot be used in psychology, for a very simple reason: in psychology, we cannot give the active principle because "the psychological intervention is always personalized" (14). Translated into medical terms, "it would be like saying that each member of the group is given a different active ingredient. We see that the experiment would simply fail. So, what works well enough in medicine cannot work in psychology" (Cagna 2019, 137; my translation). For this reason, "if there is a method of control [of the placebo/suggestion effect] in psychology, it cannot be extraclinical but must necessarily be intraclinical" (Cagna 2019, 137; my translation).

5 "The control of suggestion during the analysis never received a satisfactory synthesis and description before the MES. This is because there has never been an attempt to identify a methodological approach to make a specific and experimental discrimination between the improvements due to suggestion and those due to the truth of the construction" (Salvador 2020, 148; my translation).

6 For more on this aspect and the related theoretical debate, see Cremerius (1981).

7 This diagram is inspired by Ceschi (2020, 56).

8 This book is not intended to be an exegesis of Freud's texts or an introduction to psychoanalysis. Therefore, I do not further analyze the theme of transference herein.

9 My aim here is not to provide a general picture of Freudian metapsychology and its developments, which I take for granted. I refer instead to some important works on this subject: Farrell (1981) and Cavell (1993).

10 It is known that Freud gives two representations of this topic: the first is formed by the unconscious, preconscious, and conscious; the second is instead formed by the ego, id, and superego. It is possible to reconcile them by thinking of the conscious and the preconscious as components of the ego. However, an in-depth analysis of the two topics would take us too far from the objectives of this chapter.

11 "The facts that have caused us to believe in the dominion of the pleasure principle within the psyche also inform our assumption that one aspiration of the psychic apparatus is to keep the quantity of excitation present within it at the lowest possible level or at least to keep it constant. The latter postulate is the same as the former, albeit expressed in different terms, for if the psychic apparatus is geared to minimizing the quantity of excitation, then anything tending to increase that quantity is bound to be experienced as counter-functional, and hence unpleasurable. The pleasure principle arose out of the constancy principle; in reality, however, the constancy principle was inferred from the same facts that compelled us to postulate the pleasure principle. We shall also discover on deeper consideration that the particular aspiration we attribute to the psychic apparatus is subsumable as a special case under Fechner's principle of 'the tendency to stability,' to which he linked the sensations of pleasure and unpleasure" (Freud 2003, 80–81).

12 The definition of *form* is another problematic aspect: "The systemic organization that is responsible for interpreting the semiotic function of a sign vehicle must include a form-generating process that directly or indirectly contributes to the persistence (re-presentation) of that function. The interpretation process is constituted by generating a structure (physical form) that serves as a sign of the prior sign and also can produce further structural consequences" (Kull et al. 2009, 171).

13 A more detailed illustration of Prodi's protosemiotics can be found in Cimatti (2018, 52–58).

14 "Ever since 2001, when Sebeok asked me to review a special issue of *Semiotica* dedicated to Jacob von Uexküll, I have voiced a specific criticism to that project. I repeatedly argued that a synthesis of biology and semiotics can and should be a scientific research project where the Peirce model can be tested rather than being taken for granted. I underlined, in other words, that there are two types of biosemiotics before us: one is the extension of the Peirce's model to all living creatures, the other is a scientific approach that aims at discovering which semiotic processes actually take place in living systems" (Barbieri 2015, 167).

15 Another fundamental source of inspiration is Edelman's neural Darwinism (Edelman 1987, 1992).

16 Freud's concept of resistance is very complex, as I mention in the Introduction. We can count at least three forms of resistance. There is not only repression, which is the form of resistance exerted by the ego, but also the resistance of the compulsion to repeat (id) and the resistance deriving from the unconscious sense of guilt (superego); these are the forces that most tend to cancel out the effect of the construction truth and, therefore, to render it inoperative. Here, re-elaboration, which is the patient's work, and the pragmatics of truth mentioned in the Introduction come into play.

2 Reassembling the Mind

Psychoanalysis and Actor–Network Theory

In the Introduction and Chapter 1, we developed three theses:

1 Psychoanalysis is the only real science of the mind because it is the only one that addresses and solves the crucial problem of suggestion.
2 Psychoanalysis is based on an experimental method with its own specific logic (CWA, NaF, and CM).
3 Psychoanalysis is based on a biosemiotic conception of the mind.

My goal in this chapter is to show how Freud's approach can be strengthened through integration with ANT and the theory of material engagement (MET). This entails broadening and shifting psychoanalysis's gaze toward the active nature of material culture and the networks of nonhuman actants. However, one might wonder whether it is even possible to expand the study of the mind beyond an anthropocentric perspective. It is precisely this anthropocentric perspective that I intend to challenge in what follows.

In the previous chapter, my objective was "to take Freud seriously," showing that Freudian psychoanalysis is a science of nature that (a) works on a biosemiotic material (i.e., signs, representations, and biologically founded processes); (b) builds its own theoretical hypotheses on this material relating to the study of psychoneuroses; and (c) verifies these constructions through an experimental method based on a specific logic. My aim now is to show the need for psychoanalysis to go beyond the modern point of view and acquire a posthuman perspective. I will highlight two key aspects in this chapter: (a) psychoanalysis itself pushes us to go beyond the modern point of view (i.e., to develop a critique of the "modern constitution," using Latour's terminology) and (b) the Latourian reformulation of Freudian psychoanalysis opens up a new and unexpected scenario: that of a radical reform of the problem of the human mind, which is here understood as a collective or network of humans and nonhumans. This theoretical operation is both anti-Freudian and anti-Latourian. Specifically, Freud never posed the problem of technical reason and design; his perspective remains based on the modernist

DOI: 10.4324/9781003345572-4

predicament for which artifacts are only cold, silent objects extraneous to human beings. Using Latour's terminology, Freud thinks of the object as *Gegenstand* and not as *Ding* (Latour 2008). Latour hardly ever deals with psychoanalysis, and when he does, it is only in derogatory terms: "Psychology and its sister, psychoanalysis, think that they are rich in their infinite poverty" (Latour 1988, 205). As I will show, the renewal of psychoanalysis through ANT requires a critical analysis of both theories. Nonetheless, one point remains crucial:

> No one has ever seen a technique, and no one has ever seen a human. We only see assemblies, crises, disputes, inventions, compromises, substitutions, translations, and orderings that get more and more complicated and engage more and more elements.
>
> (Latour 1995, 6)

The ontological turn introduced by ANT allows us to recover and expand the use of the Freudian method and its logic, as well as the problem of suggestion. Through this ontological shift, the Freudian method and its logic will be applied to human and nonhuman hybrids, and psychoanalysis will become a type of reverse engineering.

Latour's Realist Constructivism

As mentioned at the end of the previous chapter, it might seem paradoxical to both affirm that psychoanalysis is a natural science and found it in Latour's ANT. At a superficial glance, in fact, Latour's approach might seem to be a sort of ultra-constructionism that tends to empty the science of its real content, with the result being a simple set of discourses and interpretations, a mere reflection of the social context. In short, there is a denial of the objectivity of science.

It is useless here to retrace the evolution of Science and Technology Studies starting from the works of Merton (1973) and the crisis of positivist epistemology (i.e., all those doctrines that define science through a break that would distinguish it from other forms of knowledge). Kuhn's (1962) work was a decisive watershed in this respect, insofar as he showed us that the evolution of scientific knowledge is moved and transformed by the succession of paradigms that are incommensurable, and not by theories defined by the simple criterion of falsifiability, using Popper's expression. To understand the crisis and substitution of paradigms, we also need to follow scientists in the non-scientific aspects of their work. After Kuhn, it became evident that the work of scientists did not take place in the empyrean, in isolation from everything else, but that it was influenced by broader social, cultural, and economic dynamics, such as affiliations, funding, power relationships, institutional relationships, and so on. In other words, the objectivity of science is full of "impurities."

Bloor (1976), Serres (1969, 1972, 1974, 1977, 1980), and Callon (1989) are the best expressions of this change of perspective that triggered a great debate in the scientific world.

Latour (1979) talks of the "social construction" of scientific facts. However, his intent is not to devalue scientific objectivity, nor is it to polemically promote relativism. Latour developed an ethnographic approach to the scientific work at the Salk Institute, which he had attended regularly. Ethnographically, Latour's approach is purely descriptive; he does not want to explain the science but to describe it while it is in action (i.e., to describe what scientists actually do, looking at them as if they were indigenous to the Amazon—with the same degree of distance and alienation). From this point of view, the construction of facts looks like a rugby match: it is a collective undertaking that brings together humans and nonhumans. Science is not an isolated theoretical enterprise; it is instead a work based on the interaction of many actors, including scientists, laboratory technicians, competing research groups, companies, financiers, politicians, public opinion, materials, cells, computers, printers, newspapers, libraries, protocols, and databases, to name but a few examples. In other words, science is a collective enterprise that also involves nonhuman agents and the mediation practices between them. Consequently, talking about the construction of facts does not at all mean denying their objectivity in the classical sense but rather showing how this objectivity is complex and presents multiple facets. Latourian constructivism is thus a realist constructivism. It is not the enemy of science; on the contrary, it tries to provide the most realistic description of science. From Latour's point of view, it is positivist epistemology that betrays the objectivity of science, reducing it to a single aspect: logic. For Latour, even though objectivity is always *adequatio rei et intellectus*, the *adequatio* is not simple—it is always mediated. It is a complex process that passes through many levels of interaction and mediation between human and nonhuman agents. To put it differently, the *adequatio* is always mediated by technical, social, and political processes.

How could we study the Amazon rainforest without the help of the *pedocomparator*, referring to a comparator case that allows us to classify the soil samples collected and identify qualities that would not be possible to identify otherwise (Latour 1999)? How would it be possible to study cells without microscopes or stars without telescopes? These artifacts do not betray the facts but translate them into measurements, calculations, representations, diagrams, and papers. They redefine them. A connection is established that takes the form of a chain between those facts (e.g., soil, cells, and stars), those artifacts, and some theories. The facts are not reduced to representations through artifacts. Instead, they are connected to the representations produced by the artifacts in such a way that the chain is always reversible; therefore, it is always possible to go from facts to tools and from tools to facts. This is the scientist's job. Can we say the same thing about psychoanalysis? Is psychoanalytic objectivity based on groups of humans and nonhumans?

Latour does not intend to stand up in favor of positivists or relativists. Instead, he seeks to understand how and under what conditions it is possible that a piece of land in the Amazon rainforest can be replaced by a number or become the protagonist of a scientific paper—a process made up of numerous mediations and translations between human and nonhuman agents, or "actants," that carry out an authentic "transubstantiation" of the Amazon rainforest (Latour 1999, 64). In Latour, *adequatio* is a synonym for translation. In this way, Latour invites the researcher to respect the fundamental principle of irreducibility when he declares, "Nothing is, by itself, either reducible or irreducible to anything else" (Latour 1988, 158). This principle establishes that no entity, however trivial, will be dismissed as mere noise in comparison with a metaphysical essence or its conditions of possibility. As Harman (2009, 13) states, "Everything will be absolutely concrete; all objects and all modes of dealing with objects will now on the same footing." Atoms and molecules are actants, as are children, raindrops, bullets, trains, politicians, and numerals. For Harman, this means that "all entities are on exactly the same ontological footing. An atom is no more real than Deutsche Bank or the 1976 Winter Olympics, even if one is likely to endure much longer than the others" (Ibid.). This is a methodological rule first introduced by Callon (1986), with enormous implications:

> Following the principle of generalized symmetry, we give ourselves a rule of the game not to change register when we pass from the technical aspects to the social aspects, hoping that the repertoire of the translation, which is in no way that of the actors studied, will convince the reader of its explanatory power.
>
> (176)

Latour invites the researcher to treat all beings equally and to place them on the same level—without reducing them to each other. To be truly symmetrical, concepts such as nature and society, or language and world, can no longer be treated as explanatory principles and instead become the problem, or what needs to be explained (see Latour and Callon 2013). They can be traced back to networks of human and nonhuman actants in constant transformation, association, translation, and power relations, in which all the actants are on the same level, and all have the same ontological dignity. We can no longer reduce them to each other by imposing hierarchical relations determined by humans.

This is the theoretical core of ANT, an unorthodox theory that has undergone multiple developments. In ANT, "we do not have individuals on one hand and technology on the other. Such a distinction would be unhelpful, because it would raise questions about the causal relationship between the two and require us to theorize and define their intrinsic attributes" (Storni 2015, 169). The contemporary world has proven itself to be an excellent ground for the application of ANT. A good example is Facebook, represented by its brand logo and name, which are forms of semiotics and communication. When anyone sees the brand logo or hears the name "Facebook," the first thing that comes to mind is the online service. The platform is accessed through technological artifacts, such as mobile and desktop devices, and

> the materiality of these artifacts makes it attractive for humans to use the service. In these devices, there are different semiotics familiar to the users that enable human interaction. These are some of the different actant interactions in the actor network of Facebook.
>
> (Williams 2020, 4)

These interactions occur as a whole, and the same thing can be said of interactions on platforms such as Instagram or WhatsApp, or of even more complex remains such as those of supply chains or transports.

Networks are dynamic, unstable, and in constant transformation. The agency of the actants is not defined based on their particular qualities, but on the basis of their position and strength in the network. According to Williams (2020, 5), "The actants are not nodes in the network; rather, the actants produce action toward one another in their interactions within the network." As Knappett (2002, 100) claims, "agency comes to be distributed across a network, inherent in the associations and relationships between entities, rather than in the entities themselves." Today, ANT is applied in the fields of banking and finance, architecture, risk management, information technology, policy studies, education, health, organizational studies, media entrepreneurship, and the social sciences (see Blok et al. 2020).

In *Irreductions* (Latour 1988)—a true anti-*Tractatus*, one of Latour's most beautiful texts—Latour states that networks are associations of human and nonhuman "actants" based on power relationships. Latour's logic is spatial in nature; there is a logical connection between the principle of *irreduction*, the idea of association and extension, and space. If entities cannot be reduced to each other, they can only associate and form networks or collectives that extend into space. Networks are therefore hybrids of nature and culture. Establishing not to reduce anything to anything else inevitably means being forced to associate entities—we cannot reduce them to a principle or to a single entity—and create hybrids. Associating without reducing means extending the network. The extension of the network is dynamic because it is based on the power relationships between the actants. Actants connect and disconnect, make alliances, and even dissolve such connections. Latour claims that actants act according to their "action programs," referring to the set of actions and reactions designed to acquire more strength in the network. Each actant "defines: what lies inside it and what outside, which other actors it will believe when it decides what belongs to it and what does not, and which kinds of trials it will use to decide whether to believe these referees" (Latour 1988, 166). Furthermore, the strength of each actant is directly proportional to its associations, since in Latour's (1988, 158) view, "There are only trials of strength, of weakness. Or, more simply, there are only trials. This is my point of departure: a verb, to try." Reality is resistance: "Whatever resists trials is real" (Ibid.). Knowledge, language, and science are nothing more than associations and tests of strength between actants. The association is a set of relations of strength and resistance in which the processes that Latour calls "translations" take place, as I mentioned before. The concept of translation is part of Latour's spatial logic. Translation, for Latour, is not the passage from one language to another but consists of (a) the definition and redefinition of equivalences, identities, and differences between actants (i.e., the boundaries of the network); (b) negotiation between different action programs; and (c) mediation between actants, or what Latour also calls "deviation."

The concept of translation is also essential in ANT developments independent of Latour's work. Williams (2020) describes translation as follows:

> A focal actant conceives an idea and draws a plan for how the idea will be fulfilled and the actants that will be involved. The focal actor then coopts these actants to support in the development of the idea. The translation process is complete when the emerging actant embodies the pattern of use. The emerging actant, be it a network or an individual actant represents what it will be used for.
>
> (6)

I summarize the main theses of *Irreductions* with this simple equation:

strength = association = resistance = translations = degrees of reality

The challenge of my research lies in applying Latourian realist constructivism to psychoanalysis—thus saving and improving psychoanalytic objectivity at the same time. In so doing, I will also refer to MET, which explores the relationship between the mind and material culture.

Taking Psychoanalysis Outside the Modern Constitution

"We might compare scientific facts to frozen fish: the cold chain that keeps them fresh must not be interrupted, however briefly," states Latour (1993, 119). The paradox of the modern Constitution for Latour is that moderns multiplicate hybrids, mixtures of nature and culture, while at the same time preventing themselves from thinking about the existence of these mixtures. From this point of view, the modern Constitution represents an enormous neurotic symptom: the repression of technical reason corresponds to an almost maniacal frenzy of reproducing artifacts, non-subjects, and non-objects—as Latour would say, the hybrids or the "third state." The modern Constitution demands purification (the absolute separation of nature and culture) while knowing its impossibility (the multiplication of hybrids); being modern means living this paradox and using it as a critical weapon against practically anything—at least until this perfect mechanism is exhausted by becoming sclerotic and consuming itself with the postmodern, which is a dead end and the admission of a defeat. Extreme deconstruction leads to dissolution:

> Those who have failed to undertake empirical studies of sciences, technologies, law, politics, economics, religion or fiction have lost the traces of Being that are distributed everywhere among beings. If, scorning empiricism, you opt out of the exact sciences, then the human sciences, then traditional philosophy, then the sciences of language, and you hunker down in your forest—then you will indeed feel a tragic loss. But what is missing is you yourself, not the world! Heidegger's epigones have converted that glaring weakness into a strength.
>
> (Latour 1993, 66)

This means being modern. The collectives are the repressed, who consist of an enormous entity that constantly puts pressure on moderns. To be modern means to endure this repression, to fight every return of the repressed, and, for this reason, believing oneself to be revolutionary. The idea of radical revolution "is the only solution the moderns have imagined explaining the emergence of the hybrids that their Constitution simultaneously forbids and allows, and to avoid another monster: the notion that things themselves have a history" (Latour 1993, 70). Moderns do what they cannot think, and they do what they cannot do. In Latour's (1993, 131) words, "How could we bring about the purification of sciences and societies at last, when the modernizers themselves are responsible for the proliferation of hybrids thanks to the very Constitution that makes them proliferate by denying their existence?" Rejecting and producing hybrids and nets are two sides of the same coin; we cannot split them. However, today, hybrids have invaded everything, and we can no longer fail to realize the work of modern repression.

From this point of view, Lacan and most contemporary philosophers are victims of the same illusion. Their illusion is based on the belief that the middle ground between nature and culture is occupied by language, and that language is strong enough to occupy the middle ground and define both poles. Lacan, and with him the psychoanalysis of object relations, remained a subject of the modern Constitution.

If we leave the modern Constitution, and if we overturn the Copernican revolution and generalize the principle of symmetry (Latour 1993, 91–100), a very different scenario immediately appears. We do not have to be either realists or constructivists; instead, we must be both. What does this mean? We need to completely redefine our conception of the mind and the relationship between the mind and the brain. We must think of the mind and brain—again, following Latour's terminology—as networks, hybrids, and collectives of humans and nonhumans. Therefore, we may quickly come to realize that our attempt to rethink psychoanalysis is not limited only to psychoanalysis but becomes something much more ambitious: a rethinking of the mind as a network of human and nonhuman hybrids. This also means showing how psychoanalysis is not just one discipline or a method among others but precisely the discovery of the mind as a collective. Is not psychoanalysis, which puts truth and error, normality, and pathology on the same footing, implicitly respecting the principle of symmetry, insofar as nothing is reduced to anything to anything else? This also means justifying the criticisms and mistrust that recur each time we talk about psychoanalysis; these criticisms and this distrust are inevitable because psychoanalysis is a challenge to the modern Constitution. Freudian psychoanalysis puts this world in crisis, as Freud broke the essence of the mind by summoning delegates, mediators, and translators, as Latour would say. This ultimately means rethinking ANT itself.

From this point of view, I intend to connect the three crucial aspects of my investigation in this Chapter: psychoanalysis as an experimental method, psychoanalysis as biosemiotics, and psychoanalysis as a study of the construction of the mind as an actor network. I will argue that there is one thread that unites these three aspects.

If being nonmodern means that "we do not need to attach our explanations to the two pure forms known as the Object or Subject/Society, because these are, on the contrary, partial and purified results of the central practice that is our sole concern" (Latour 1993, 79), then being nonmodern in psychoanalysis will mean following the construction of the mind as an actor network. In other words, to think of the mind as a collective is to redefine the old question of mind–brain dualism. This dualism dissolves and disappears when, following Latour, we conceive of the mind and the brain not as essences but as existences (i.e., as the result of the splitting of human and nonhuman networks). It is important that this first point be grasped: the mind is the provisional result of the association between beings and their power relationships. The mind, to quote Latour again, has nothing mental because it is only a certain type of association between humans and nonhumans. How do we distinguish this association, or network, from others? I think Latour does not provide the tools to answer this question. In fact, Latour is the victim of a paradox: while trying to think of collectives, he deprived himself of the adequate means to do so because he based his analysis on Greimas's semiotics, a single semiotic model. Instead, I propose that we maintain the methodological premises of ANT, but found them on Peirce's semiotics and biosemiotics, a different basis that provides a much more flexible model—in the sense that it admits many more variations—and which is not as anthropomorphic as the Greimas version. I will develop this point in the next section.

In doing this, I restrict my analysis to a single type of collective, or what I call bio-collectives, referring to those collectives to which the principles and tools of biosemiotic analysis can be applied. Therefore, I do not extend biosemiotics to the study of all collectives, but only to a part of them: the bio-collectives. Bio-collectives are composed of humans and nonhuman living beings (e.g., bacteria, viruses, cells, animals, plants, and fungi) in a nonhuman living world much larger and more complex than the human one. The mind is a bio-collective; this is my thesis.

Latour and Semiotics

I share with Latour the idea that an approach to the study of networks must be empirical and semiotic, but I propose improving his approach by introducing a different semiotic model. In so doing, I specifically follow Kohn's "anthropology of the beyond-human" (2013), which translates Peirce and Deacon's theses into ethnography.

The semiotic method is, for Latour and ANT in general, an indispensable model for understanding the sense of technology and the social because it allows us to criticize the modern split between matter and meaning, and between the real world and the symbolic world. The entities of the world act and produce signs and meanings regardless of their anthropomorphic nature. This explains the role and function of hybrids. According to Latour, semiotics is inherent in the concept of the network; the actant acquires a role and a value not for an alleged essence, or a set of intrinsic qualities, but for the semiotic relationships it has with other actants.

Now, Latour uses a Greimas-inspired model of semiotics. In my view, this model is not only incompatible with the principle of human/nonhuman symmetry but also involves the risk of narrativizing the networks and does not explain the material origin of the meaning. I am not claiming that Latour uses the Greimas model in a non-critical way. Latour is aware of the risks inherent in the model. However, this does not mean that all risks are eliminated.

Fully respecting the principle of symmetry would imply the complete abandonment of the Greimasian narrative model because it is based on an anthropocentric vision of meaning, for which human language—and above all narrative—is the fundamental model in the construction of meaning. Greimasian semiotics is based on an interest in identifying the type of role of an actant based on the relationship it establishes with the human actor (Ventura Bordenca 2021, 18–19). For Greimas (1987), the semiotic universe coincides with human culture. The key elements of Greimas's semiotics of discourse are the procedures of narratology and the concept of narrativity. Greimas's semiotics is generative and transformational, and two levels of narrativity are distinguished: the surface level and the deep level. The first level is that of the linguistic structures of the story, or the stories actually produced by oral and written traditions (a level also called figurative), while the second includes the structures common to all narratives and is of a conceptual, logical, and nondiscursive nature. The goal of Greimasian semantics is to demonstrate the plausibility of deducing, through intermediate and homogeneous levels, the superficial structures of the story with deep structures (i.e., the possibility of deriving the narrative in progress from some initial semantic conditions with a simple algebraic calculation). The most elementary structure is the "semiotic square," which is the set of the simplest logical relations that make a *quid* meaningful: contradiction, opposition, and presupposition. In other words, if something has meaning, it is not because we somehow guess what it means, but because we can frame it within a system of logical relations.

From the semiotic square, Greimas (1987, Chapters 4–5) deduces the fundamental grammar of narrative language, which, like any grammar, has two components: a morphology (i.e., a taxonomy that fixes the fundamental terms of the grammar) and a syntax (i.e., the system of operational rules to manipulate and use the terms). The first contains the relations of the square and conceptualizes them as static categories; it therefore illustrates a formal, achronic model that is a pure taxonomy of terms. The second "consists of operations carried out on terms that can be invested with content values; the syntax thus transforms and manipulates these terms by negating or affirming them, or, and it amounts to the same thing, disjoining and conjoining them" (Greimas 1987, 70). Syntactic operations, since they are within the established taxonomic framework, "are oriented, and thus predictable and calculable" (Ibid.). With the passage from morphology to syntax, the first narrativization of the model takes place. The same logical relationship in the semiotic square is conceived of in the taxonomy as a category of the narrative, while in the syntax, it is understood to be an operation or an action. For example, the contradiction between two terms becomes the operation of negating one of the terms and simultaneously affirming the contradictory term. Therefore, in Greimas's view, the surface level emerges from the application of the anthropomorphic notion of doing to the basic logical operations in the semiotic square. "Establishing the equivalence between the operation and the doing introduces the anthropomorphic dimension into the grammar," states Greimas (1987, 71). It is then possible to define a "canonical narrative scheme" that entails "the repetition of three tests—qualifying, decisive, and glorifying which appeared as the regularity, located upon the syntagmatic ax" (Greimas and Courtes 1979, 204). The "actant" is a syntactic unit of the narrative that acquires the ability to act only in the set of relationships in which it is inserted. In this sense, "an actant can be thought of as that which accomplishes or undergoes an act, independently of all other determinations" (Greimas and Courtes 1979, 5; see Greimas 1987, Chapter 6).

This Greimasian perspective just outlined greatly influenced Latour, as Akrich and Latour (1992) demonstrate. In this important text, meaning is defined in a very general way: "the word meaning is taken in its original nontextual and nonlinguistic interpretation; how one privileged trajectory is built, out of an indefinite number of possibilities" (Akrich and Latour 1992, 259). Semiotics is the study of the construction of a trajectory and can be applied to chains of humans and nonhumans, all considered actants (i.e., poles on a continuum upon which agency is distributed); "what the analyst is faced with are assemblies of humans and nonhuman actants where the competences and performances are distributed; the object of analysis is called a setting or a setup" (Akrich and Latour 1992, 259). The goal of the analysis—says Latour—is to follow the construction of meaning in the device, and it is precisely here that the influence of textual and narrative models emerges. The analysis must define what Latour calls the "script" of the device:

> The aim of the academic written analysis of a setting is to put on paper the text of what the various actors in the settings are doing to one another; the de-scription, usually by the analyst, is the opposite movement of the in-scription by the engineer, inventor, manufacturer, or designer.
>
> (Akrich and Latour 1992, 259)

The influence of the textual and narrative models is confirmed by the concept of the "action program," which is of crucial importance for Latour (and it will be for us, too, in the next chapter). In fact, the script is the set of action programs of an actant and their transformations. The action program "is a generalization of the narrative program used to describe texts, but with this crucial difference that any part of the action may be shifted to different matters" (Akrich and Latour 1992, 260). In the semiotic analysis of artifacts, the narrative of the transformation of action programs must also include the intervention of nonhumans, since "the aim of the description of a setting is to write down the program of actions and the complete list of substitutions it entails and not only the narrative program that would transform a machine in a text" (Akrich and Latour 1992, 260–61). It seems that, for Latour, the passage from artifacts to their description (i.e., the script) is narrative, while the reverse passage from the description to the artifacts is not. Nevertheless, many analyses in the field of semiotics (Landowski and Marrone 2007) or hermeneutics

(Coeckelberg and Reijers 2020) show the opposite, insofar as technologies are held to be "materialized narratives." Coeckelberg et al. (2021) also demonstrate the possibility of applying other models of narrative analysis to technology, such as Ricoeur's hermeneutics (Ricoeur 1983, 1985).

As Lenoir (1994) points out, it is difficult "to find [in Latour] a satisfying discussion of how we get from Greimas's world of texts and narratives to the world of collective entities, quasi-objects, and nature-culture." In Latour, the set of possibilities through which we define meaning is predetermined by a fixed grid (i.e., the canonical narrative scheme). This involves a double risk of idealizing the text and its counterpart, which would be naive realism, as these are two opposite poles between which Latour oscillates. Is the idea of network, then, a metaphor? Is it a model, or just a narrative? What kind of explanation does it provide? On the one hand, we find the risk of an extreme deconstructionism for which everything is resolved in narratives and interpretations of narratives; on the other, we have a naive realism that does not know how to differentiate between the model and reality or between different models.

Latour falls into the "sin" of anthropocentrism despite his defense of symmetry; this is because the source of the associative bond is always the human being and its narrative ability (i.e., the canonical narrative scheme). Nonhumans need human spokespersons to speak and act in this narrative. Without human intervention, meaning cannot flow into networks. This Latourian tendency to narrativize the networks can also be seen in the style of many other texts, such as Latour (1993a), where the protagonist is Gaston Lagaffe, a famous cartoon character created by the designer André Franquin; Latour (1993b), which is dedicated to the Berlin key; Latour (1993c), which instead reports on the daily life of the Center d'Histoire des Sciences in Les Halles aux Cuirs, starting from a door; and also *Irreductions* (Latour 1988), which starts with the story of Robinson Crusoe. It seems that we cannot think of the network without narrativizing it (i.e., "bending" it to human initiative). Still, in my view, this would be a betrayal of the principle of symmetry, as "the inscription of constructors and users in a mechanism is very similar to that between authors and readers in a story" (Latour 1993a, 67; my translation).

This point is connected to two others that are equally important. First, Latour fails to explain the material origin of meaning. While seeking to transcend the dualism between matter and meaning, he always keeps himself on a single semiotic level, the narrative level, and tries to explain all the collectives in this way. The implicit anthropomorphism in this view obstructs his understanding of how meaning comes from matter. The use of biosemiotics can help overcome this flaw by explaining the origin of meaning and language, starting from the semiosis *within* living matter. In other words, Latour is unable to explain the "material basis of meaning," quoting Prodi. For this reason, he remains profoundly modern.

The second point is closely related to the first. In Latour, nonhumans are always machines; such entities include the laboratory, the microscope, the pneumatic pump, the Berlin key, the door of a research center, and even an innovative transport system. Latour analyzes artifacts and technologies, but never animals, bacteria, or plants; he lacks "the basic intentionality of the pig" (Kohn 2013, 160). When he tries to do this, as in Latour and Strum (1986, 1987), baboons are treated as machines. Latour (1988) reduces the microbe to an actant in Pasteur's script—the microbe does not speak, so it needs a spokesperson, and it does not speak because the fundamental semiotic model for Latour is human language. In fact, the microbe cannot speak, and this is a great limitation for Latour. The reason for this shortcoming is that living nonhumans are not as easily reducible to the narrative and textual model as nonliving nonhumans are. Greimas's anthropocentrism and textualism make Latour unable to truly think of the nonhuman—not only machines but also nonhuman living beings. This is an important point because it reveals a "dark side" of the principle of symmetry. Electing not to reduce these entities does not mean ignoring their differences and, therefore, confusing or conflating everything about them. I claim that networks are not all the same, and saying this does not mean imposing a hierarchy. Not all networks can be reduced to text and story models; in other words, there are networks that cannot be told. This is one of the central theses of this book; there are remains that cannot be translated into a text, and for which we need a different model of semiotics. These are *unconscious networks*.

What do I mean by saying, "There are networks that cannot be told"? I claim that there are networks to which it is not possible to apply the canonical narrative scheme. The first reason is that the concept of association is not necessarily logical and cannot be reduced to a single univocal scheme. Perhaps there should not be a call for a single scheme for all networks, since there are forms of association and networks that are not cultural and still able to produce meaning. For instance, a cell is a network made up of nonhuman agents that produce meaning (i.e., it interprets a code, the DNA, and makes choices in its environment). Moreover, there are networks that, while respecting the semiotic square, cannot be translated into the canonical narrative scheme. An example is WhatsApp, which is a network of human and nonhuman actants. Can we reduce WhatsApp to a single narrative? Doing so would mean giving an anthropomorphic interpretation of the network and therefore preventing us from grasping other important aspects. Can an algorithm always be reduced to a narrative? No, not always. Can the relationship between human designers and algorithms always be reducible to a narrative? No, not always.

This point clearly emerges from Latour's analysis of Kohn (2013). Based on four years of research with the Runa population of the Upper Amazon in Ecuador, Kohn (2013) redefined the work of anthropology by showing how the traditional Western point of view collapses when confronted with the "thinking of forests." We cannot understand the Runas and their relationships with the surrounding environment if we do not consider the forms of nonhuman communication, the "living semiosis," with which the Runas are constantly in contact. The forest ecosystem thinks because it is capable of representation, as it produces and interprets signs and therefore acts on them. Kohn conceives of living semiosis in Peircean terms (i.e., as a network of icons, indexes, and symbols that evolve according to an emergency logic); this type of approach has also been described by Deacon (1997). The general idea is that the nonhuman world (e.g., plants, animals, fungi, and bacteria) is capable of semiosis and representation and that humans have lost the art of listening to this world, as well as the ability to establish "ecological relationships."

In his review of Kohn (2013), Latour formulates many criticisms. First, he asks, "Why is it presently better to handle connections among entities through semiosis rather than through associations?" (Latour 2014, 263). It is not clear why one thing excludes the other. In other words, semiosis implies an association. Association is a semiotic process because it uses signs and is an elementary structure of life. Sure, "when you study animists (in the new sense given to the word by Descola [2013]), it is less difficult to 'animate' entities well 'beyond' human souls" (Latour 2014, 263). However, this is precisely the radical challenge of an ANT based also on biosemiotics and not on Greimas. Latour (2014, 264) adds, "In spite of what Kohn often says, automorphisms do not define the background of the world as it is, but only one register of how to handle connections by treating all of them as being relations among selves." However, Peircean semiosis does not rule out other, more complex forms of association.

Latour recognizes that "Greimas could have difficulties making ontological claims," even though "he can entertain a vast diversity of registers" (Latour 2014, 64). In other words, according to Latour, whereas Greimas allows for a plurality of semiotics, and therefore of collectives, Peirce does not. There would exist, in his eyes, only one semiotics of Peirce. Here is the fundamental criticism: "Peirce allows strong ontological claims but has to stabilize much too fast all connections into automorphisms" (Latour 2014, 64). Further on, Latour (2014, 64) maintains that Peirce's semiotics "claims to be an alternative description of what the world is." Therefore, if Greimas's model allows for the plurality and dynamism of collectives to be maintained by virtue of the multiplicity of its semiotics, Peirce loses all this by stabilizing the collectives too much. For this reason, Latour (2014, 64) denounces "the danger of stabilizing too quickly what the furniture of the world is, and the necessity of having a semiotic toolkit able to restart the negotiation whenever it has stalled."

Latour's position can be challenged in several ways. First, Latour defines Peirce as "a fairly cryptic philosopher" (Latour 2014, 62), ignoring—or unwilling to consider—all the incredible developments that Peirce's semiotics has led to in the last half century. He makes no reference to the application of Peirce's semiotics in biosemiotics, zoosemiotics, cybersemiotics, or cinema and image theory, for example. The endless classifications of signs in Peirce also confirm the plasticity of the Peircean approach. Furthermore, in Kohn's (2013) perspective, there are not actants but "places" of semiosis, a concept that is even more mobile and plastic than that of the actant. Kohn (2013) explicitly says that the "places" of semiosis are provisional and variable. Indeed, everything can be a "place" of infinite semiosis. The accusation of a lack of variety is therefore inconsistent. Instead, the opposite is true; in Greimas, there is a great variability of semiotics, but all these refer to a single narrative model and pre-established scheme. Second, even the accusation of stabilizing the collectives crumbles in the face of the evidence. The evidence concerns, in this case, the nature of Peirce's semiotics, and the principle of infinite semiosis is profoundly dynamic.

Third, Latour's claim that Peirce's semiotics "claims to be an alternative description of what the world is" can also be challenged in several ways. It is true that Peirce identifies semiosis and reality. However, no matter how systematic Peirce's philosophy is, we are not forced to keep ontology and semiotics together. Once again, the developments in biosemiotics clearly demonstrate that we can use Peirce's semiotic analyses without implicitly assuming their ontology. From this point of view, Peirce is much more relativist (in a Latourian sense) than Greimas. Fourth, in Peirce, there is no logical presupposition of semiotics, like in Greimas, but semiotics is based on habits, which are empirical regularities, not conceptual rules. Kohn (2013, 280–283) also mentions "forms," which are self-organized and dynamic structures; these forms are connected to habits. Fifth, Latour is right in claiming that "no matter how good Kohn's book is, the Runa qua Peircian ontology have not become for everybody else the definition of their common world" (Latour 2014, 64). However, he is mistaken in assuming that this is Kohn's aim.

There is also a sixth point that I would like to highlight. The definition of force in Latour goes very well, paradoxically, with the concept of a sign in Peirce:

> No actant is so weak that it cannot enlist another. Then, the two join together and become one for a third actant, which they can therefore move more easily. An eddy is formed, and it grows by becoming many others.
>
> (Latour 1988, 159)

The sign is connected to the object, and in this way, through this association, it confronts a third party, the interpretant, who acts on them and transforms them. The sign is a collective. I will develop this point further in the next section.

In Latour, paradoxically, there is too much ontology and too much semiotics, but too little empiricism. The Peirce–Kohn model explains networks through biosemiosis, and in this way, it gives the networks a connection to reality at the same time. This connection to reality is an empirical basis that the Latour–Greimas model loses; it is also a dynamism due to the infinite process of semiosis.

My claim is that the Greimasian model should be replaced by a Peirce-inspired perspective. I maintain the main conditions of the ANT and the ontology of *Irreductions*, but place as a basis not Greimas, but Peirce. The simplest collective model is the sign in the Peircean sense (for a complete analysis of Peirce's semiotics, see Chapter 4). This semiotic basis is more flexible and dynamic than that of Greimas. Following Peirce, therefore, I affirm that the sign is not an abstract human cultural construction but a real structure that arises from the organization of matter. Many other semiotic models are grafted onto this basic structure, ranging in diversity from biosemiotics to cultural semiotics. What I mean is that Peirce can be used as the basis for an interactive dynamic of different semiotic models chosen according to the network we are studying.

ANT + Biosemiotics = Bio-Collectives

Biosemiotics gives us the tools to solve some intrinsic problems of ANT:

1 How are associations formed?
2 How are the boundaries between the networks established?

I intend to apply the methodological premises of ANT to the study of biological systems (i.e., living organisms) using a Peircean approach to biosemiotics. The conditions for the composition of the network and the admissibility of its elements are defined by biosemiotic relationships. The unity and survival of the collective are contingent upon the ability of its members to produce and interpret signs. The translation process is nothing other than the passage of the semiotic process from one actant to another. In other words, to guarantee the unity of the collective, each actant must be the *interpretant* of a previous semiotic process. I call a bio-collective an association of human and nonhuman living actants based on biosemiotic relationships. I will therefore focus my analysis on the study of bio-collectives. The main advantage of this position is the overcoming of the meaning/matter dualism; for biosemiotics, the sign and the meaning arise from the organization of matter.

We can now answer the aforementioned questions:

1 Networks are formed starting from biosemiotic conditions; the associability of x to the network y is established by the ability of x to interpret that relationship (e.g., the association in that network) and translate it.

2 The boundaries of the network are the boundaries of the umwelts of its actants.

As Hoffmeyer (1996, 20–22) shows, there are some fundamental semiobiological processes, such as embryogenesis and reproduction. In both cases, DNA is a sign that is interpreted in different ways by internal agents of the cell. The greater specialization of eukaryotes compared to prokaryotes has allowed for the development of not only a vertical semiosis (the interpretation of the genetic code over time and inheritance across generations) but also a horizontal semiosis in space. DNA is transmitted almost exclusively to the next generation. According to Hoffmeyer (1996, 32),

> There is a tradition in biology for depicting time as a vertical axis (cf. the word descent), and it can therefore be said that the eukaryotic organism translates genetic communication into a purely vertical phenomenon, transmitting from parents to progeny—in other words, a *vertical semiosis*.

To counteract this "privatization" of the genetic material, "the eukaryotic cells have, however, developed ingenious and efficient methods of communicating with one another by chemical means, primarily through physical contact" (Ibid.). Special proteins on the surface of eukaryotic cells "can, so to speak, poke their noses into their neighbors' affairs" (Ibid.). In other words, a means of communication evolves that is not based on signs in the form of genes but on signs in the form of proteins or other types of chemical compounds. With respect to this means, "One could call it *horizontal semiosis*, the exchange of signs through the three dimensions of space rather than through time. Not so much genealogical semiosis as ecological semiosis" (Ibid.). The combination of vertical semiosis and horizontal semiosis constitutes the umwelt of a living being. For Hoffmeyer (1996, 58), "The specific character of its umwelt allows the creature to become a part of the semiotic network found in that particular ecosystem. It becomes part of a worldwide horizontal semiosis." Hoffmeyer's theses are very close to those advanced by Deacon (2011, 274–275) and to his distinction between *homeodynamics*, *morphodynamics*, and *teleodynamics* (see Chapter 1).

As we can see from these simple examples, there are some initial material conditions from which—as Prodi argued (see Chapter 1)—greater complexity and semiotic relationships develop; these are based on operations of interpretation of a basic code (e.g., the DNA) in relation to a context. Based on these vertical and horizontal semiotic relations, collectives are established. In fact, what Hoffmeyer (1996, 35) calls habituation, a concept taken up by Peirce, is connected to semiosis. Habituation is the tendency of nature to acquire clothes, regularities, and trends. Within this context, physico-chemical habits became biological habits. "Primitive cells were organized into endosymbiotic patterns which we call eukaryotic cells," and then "eukaryotic cells acquired the habit of working together as multicellular organisms which in the course of time adapted to the prevailing logic of the ecosystems" (Ibid.). The stabilization of these cellular conditions, which are semiotic, has allowed the development of increasingly complex forms of life, such as animals, following the logic of emergency. Having an ever more sophisticated umwelt is a great advantage because it allows one to manage the absence, in this case, the future, in an ever more effective way (see Deacon 1997, 2011). The set of umwelts is what I call a *semiosphere-biosphere*.

The first objection to these theses is that studying bio-collectives is futile when attempting to understand the large networks in the contemporary world, which are neither biological nor biosemiotic. For instance, it may be argued that WhatsApp is not a bio-collective. However, this objection suffers from modern prejudice. The criticism does not take into account the fact that my thesis is not general; I am not saying that all collectives are bio-collectives—or that they can be reduced to bio-collectives. Instead, I claim that the current ecological crisis and the emergence of Gaia show that ignoring bio-collectives leads to the downfall of all collectives (more on this in Chapter 6). From this point of view, Gaia is the big "biosemiotic unconscious," that is, the repressed of the Anthropocene. It is, therefore, Gaia who—as Latour suggests—we must face in order to avoid really going crazy. Therefore, WhatsApp in itself is not a bio-collective, but it can become part of the bio-collective if it is able to meet the admissibility conditions. Nothing prevents nonhuman nonliving beings from becoming members of a bio-collective.

The second objection concerns whether we can delimit the boundaries of bio-collectives. We can delimit these boundaries if we start from the observation of how the different actants define and organize their umwelts. The conditions of the power relationships, the intensity of forces, and resistances are biosemiotics (i.e., they depend on the way in which the different actants build their umwelts).

Let us take an example provided by Latour: that of the "battle" between the Mississippi and the lesser-known river, the Atchafalaya, which flows below the first and threatens to invade it (Latour 2017, 30–35). The actants in this network are the two rivers, the US corps of engineers, and the dam built to divide the two rivers and avoid massive flooding. The power relations between these actants depend on their respective umwelts (i.e., on how the rivers, the corps of engineers, and the dam build their umwelt from certain material conditions). The river is an actant that bases its action on the initial material conditions and selection processes based on a logic of importance and emergency. The dam also has an umwelt, which is a deviation-extension of the umwelt of human actants, who are in this case the engineers. When I say that Mississippi has an umwelt,

I mean that it implements different kinds of semiotic relationships. The Mississippi is a collective (including the composition of the seabed, temperature, lithology and geomorphology of the hydrographic basin, type of water, flora, fauna, and dynamic processes that regulate the exchange of energy and matter with the outside), and the actants that compose it have semiotic relations; for example, a certain type of seabed favors the development of a certain fauna, and this function is not present in the physical structure of the seabed but is born in the relationship with that fauna; that type of seabed is thus a sign of the presence of that fauna.

In addition, the Mississippi collective made a series of selections, starting with certain initial material conditions. The choices made are significant because they are important to the Mississippi collective, that is, to its composition, structure, and equilibrium. They are also signs of the initial material conditions, as they refer to them even in their absence (i.e., when they disappeared). The choices made by the river refer to material conditions not because they try to adapt to them, but because they try to shape and build their own umwelt. The Mississippi collective defines its own umwelt and therefore its future (i.e., the unity and survival of the collective itself). Now, one is to imagine, as Latour (2017, 33) explains, that this umwelt collides with that of the Atchafalaya River. However, this collision would be incomprehensible if we did not consider biosemiotic networks. Furthermore, it would be incomprehensible if we did not consider the relationship between these umwelts and the human umwelt, with its own interests and values. The collision between the two rivers would not be a disaster and would not require the construction of a dam if it did not jeopardize the American economy. The dam was built to save the unity of the semiosphere-biosphere, that is, the coexistence of all umwelts.

Maintaining the methodological principles of ANT requires us to place all beings on the same footing and, therefore, use the same register for all of them. We are thus forced to admit that even the nonliving actants in the bio-collective have an umwelt, insofar as they start from certain material conditions, interact with the surrounding environment, and produce meanings. In other words, the umwelt of nonliving nonhuman actants consists of deviations or translations of the umwelt of living humans and nonhumans. The artifact is part of the bio-collective insofar as it maintains the unity of the bio-collective. We forget that nonliving actants are not entities extraneous to living semiosis; they affect and are affected by living semiosis. They become capable of producing meaning from living semiosis. In understanding these processes, we respect the principle of symmetry.

Therefore, in the case of bio-collectives, we can then reformulate the Latourian equation as follows:

> umwelt (strength = association = resistances = translations = degrees of reality)

The Biological Genesis of Technology

There is a point that we have not dealt with sufficiently so far but that is essential: How is technology born within the bio-collective? In other words, how can biosemiosis produce an artifact? The conceptual resources needed to answer this question are offered to us not so much by biosemiotics or ANT as by the French philosophy of technology. Therefore, let us go back to the starting point of our investigation: Stiegler's organology.

As I explained in the Introduction, for Stiegler, technology is organized inorganic matter that arises from a process of externalization from organic matter. Technology is a product of biological life that transforms biological life itself through evolution recursively. Evolution, in fact, is like a wheel that turns on itself. Tertiary retentions, the result of primary and secondary retentions, constitute the individual and collective umwelt and then modify, in turn, the primary and secondary retentions. Now, the semiotic relationship, as we have described it so far, can be considered a form of tertiary retention in the Stieglerian sense. The sign sets a material organization of life based on the relationship with absence. The sign is already a technics, or rather, a mnemotechnic. The movement that goes from the sign to the interpretant is a retention; the interpretant "recovers" the sign. Stiegler himself supports this thesis: mnestic milieux are also symbolic milieux (2006, 55).

The integration of Stiegler's point of view of organology with biosemiotics allows us to answer a key question: how to explain the genesis of the machine in biological and organic terms. I claim that technology is an epiphylogenetic memory mediated by biological semiosis and by the logic of emergence that characterizes it. Without this mediation, it is impossible to understand technology. The externalization is therefore, above all, a semiotic process, involving the production and interpretation of signs (i.e., icon, index, and/or symbol). The machine is always the result of a semiotic operation within an umwelt and is therefore the result of an intrinsic tendency of biological matter. This is not a deterministic solution; as we have seen, semiotic processes and the codes that regulate them in biological life are not fixed and immutable laws—they admit variation and creativity. Artifacts, machines, and technical systems are therefore networks of signs that are rooted in biological life.

In other words, if biosemiotics (i.e., natural signs and codes) is what the human being uses to build their umwelt, then technology must necessarily be based on biosemiotics if it is (1) an externalization of life, (2) a mediation between the human being and the surrounding environment, and (3) an instrument capable of adapting to the environment and to the human being. In this scheme, however, as Simondon (1965) points out, a key element is missing: (4) invention, understood both in a biological sense (a solution to a problem) and in an ontological sense (such as ontogenesis, which is the creation of new signs and codes and a new umwelt). "The technical operation requires a technical and natural life" (Simondon 2005, 175; my translation). However, technology adapts to the human umwelt by transforming it recursively. The invention produces a completely new type of being, the result of a design process, and endowed with an organizational autonomy similar to that of living beings. However, do these conditions apply to any type of technology? Does an artifact such as software have an umwelt? Can we say that software builds its own umwelt by collaborating with other artifacts and modifying the human umwelt? Nevertheless, we are used to thinking that a computational system is purely syntactic; as such, how can it produce semiotic relations, meaning icons, indexes, and symbols capable of building an umwelt and transforming the human umwelt? I will elaborate on these questions in Chapter 5. For the moment, I want to focus on the relationship between technology, life, and biosemiotics.

The tendency of life toward technicization and the analogy between the artifact and living being are important concepts in the French philosophy of technics. Canguilhem (1966) develops a biological philosophy of technics in which technics is understood as a manifestation of life. The central concept is that of "biological normativity." Life, Canguilhem claims, is a normative activity—that is, it involves the choice and definition of values. It is an idea very close to Prodi's: the living organism, guided by its vital needs, differentiates what is in front of it; in other words, it qualifies something as significant and something as non-significant. Drawing upon Bergson, Canguilhem states, "Life is not just submission to the context but the institution of one's own context. In this way, life places values not only in the context but also in the individual itself. This is what we call biological normativity" (Canguilhem 1966, 155; my translation). Life "is defined as a normally creative activity, in the precise sense that the normal living is the one who invents new rules of life, who has a creative relationship with the context" (Clarizio 2021, 28; my translation).

Criticizing neo-positivism, Canguilhem affirms the anteriority of technology with respect to knowledge and science. This leads him to affirm that technology is the manifestation of a "creative power," preceding any knowledge or science, which has its roots in the biological life of humans (Canguilhem 2011). However, according to Canguilhem, this does not mean supporting a pragmatist position, meaning "a philosophy that attempts to reduce all functions to the functions of vital, utilitarian or technical adaptation" (Canguilhem 2011, 501; my translation). It is necessary, he affirms, to avoid two parallel risks: reducing technology to an application of science (i.e., neo-positivism) or reducing science to a theorization of technology (i.e., pragmatism).

In *Le normal et le pathologique* (1966), Canguilhem defines the terms of a biological genesis of technology. He affirms that the independence of science and technology does not prevent their interdependence, as shown by the case of medicine, which is both technology and science. In the case of medicine, the pathological precedes the normal. There is first a human need to heal and then the will to find ways to overcome the disease. This will is the expression of the creative power that belongs to the living. From this perspective, technics is considered a manifestation of biological life. This manifestation should be understood not as a simple extension but as a response to the needs of life itself. Technics is not a manifestation of the spirit but a pre-theoretical, unreflected, entirely biological activity: "It is an activity that is rooted in the spontaneous effort of the living to dominate the context and organize it according to living values" (Canguilhem 1966, 156; my translation). The normativity of life is inseparable from a technical transformation of the context; life is always "information activity and the assimilation of matter" (Canguilhem 1966, 80; my translation). Technics is therefore what guarantees the continuity between life and knowledge. It is science that creates a break between technics and life.

The centrality of the relation between life and technics in Canguilhem's philosophy is also clear in another important paper, the "Note sur la situation faite en France à la philosophie biologique" (2015). In this paper, Canguilhem directly criticizes a mechanistic view of biological life. This vision, he claims, cannot explain the genetic and dynamic phenomena intrinsic to life. In the paper, Canguilhem uses the expression "organologie générale," indicating a philosophy of biology capable of showing the continuity between life and technics, or "machines as organs of life" (319; my translation). Technology is "a strategy that life has always used to impose itself in the fight against the context" (Clarizio 2021, 30; my translation). The biological organ produces the technical tool, the machine, and that is why the machine can be used analogically as a model for the knowledge of organs, as Descartes did.

However, the price to pay for this thesis is the difficulty of recognizing the originality and ontological specificity of technology. This is the problem that emerges in another text, *Machine et organisme* (Canguilhem 1965), in which machines are reduced to mere bunches of independent parts whose synthesis and effectiveness depend only on humans. This is explained by a methodological choice; the purpose of Canguilhem is never to understand the essence of the machine "but to show the vital origin of mechanistic theories" (Clarizio 2021, 33; my translation). His philosophy of technics is only an appendix of his epistemological reflection on the living and on the organization of society. It is Simondon who develops a proper philosophical analysis of the ontological originality of machines.

Simondon (2005) links technology to the question of individuation, which is an ontological dynamic. He also aligns very closely with Bergson, whose reflection on technology is an integral part of his reflection on life and biology (Barthélémy 2005). According to Simondon, technology is born as a response to a problem that arises in the relationship between the organism and its vital context:

> What situation does the invention respond to? To a problem, or to the interruption by an obstacle, by a discontinuity that blocks the way. . . . The invention is the appearance of the extrinsic compatibility between the environment and the organism and of the intrinsic compatibility between the subsets of actions.
>
> (Simondon 2014, 139; my translation)

In this passage, it is clear to what extent the Bergsonian conception of life as a relationship with matter is present in Simondon

> and how in the two authors we find the same idea of technical life, or the idea of life as schematism between organism and the environment by means of an inventive technical activity, of which the technical object is only the most complete form because it is objective.
>
> (Clarizio 2021, 167; my translation)

By *schematism*, a concept also present in Bergson's essay *Matière et mémoire* (1965), we do not mean the idea of the Kantian mediation between intuitions and concepts but the search for solutions for concrete problems by examining the compatibility between image and reality. At the root of technics, there is, therefore, the vital dynamics of invention. Technical schematism is a true biological function because "it involves transferring the relational character of biological individuality to the level of creative invention" (Clarizio 2021, 180; my translation).

Stiegler (2006, 65–66) develops two fundamental criticisms of Simondon: (1) he does not speak of technical individuation, even if he constantly speaks of technical individuals, and (2) he never talks about the role of technical identification in the relationship between psychic individuation and collective individuation, that is, the dimension of the transindividual. Simondon, from Stiegler's point of view, remains a prisoner of the traditional dualisms of subject/object and culture/nature. For Simondon, according to Stiegler, technics is only a moment of psychic and collective identification. On the contrary, Stiegler states that technics is the fundamental condition of individuation because it allows the passage from psychic to social individuation. In other words, technical objects are epiphylogenetic memories that form the pre-individual level of psychosocial identification. However, this is a pre-individual level that is continually defined and renewed in relation to the forms of psychosocial individuation; epiphylogenetic memories are the tools through which life is defined and renewed.

ANT + Biosemiotics + Psychoanalysis = The Mind as Bio-Collective

For Latour (1993a, 114; my translation), "The beauty of artifacts lies in the fact that they take upon them the contradictory desires or needs of humans and nonhumans." If we take these words seriously, then our project will be both clarified and enormously complicated at the same time. The passage through Latour helps us to complete the enterprise begun in Chapter 1, as we can now re-evaluate Freudian psychoanalysis as a science of nature based on a specific experimental method. In so doing, we will come to realize that psychoanalysis becomes the testing ground for a radical reorganization of the way we think about the human mind. In this section, I want to develop a reinterpretation of the psychoanalytic method and mental model through ANT and biosemiotics.

I advance two preliminary considerations that I will justify in the next section. The first is that Freud pushes us to think in this nonmodern direction. The second is that, if we admit the first consideration, then we realize that there is a thread that connects psychoanalysis and AI.

An important contribution to understanding the mind as a collective comes from the MET, which is compatible with Latour's ANT. The MET is an essential point of reference in my work, even if—in my view—it remains too intimately tied to a "neurocentric" perspective and other unclear concepts, such as that of "mental space" (Malafouris 2013, 103). Reinterpreting some aspects of the MET from the point of view of Peirce's semiotics can strengthen the theory; for example, the concept of cognitive projection can be perfectly reinterpreted in terms of Peirce's infinite semiosis; indeed, this reinterpretation reinforces it. Furthermore, the MET allows for the construction of some important connections between psychoanalysis and the cognitive sciences.

I now establish some fundamental actants of the mind, along with four assumptions:

- Intelligence (set of the patient's cognitive abilities)
- Psychoneurosis (set of resistances)
- Human body (organic stimuli)
- Material signs
- Linguistic signs
- Objects
- Drives
- Constructions of the analyst
- Analyst's interpretations

Four assumptions:

1 The condition of admissibility in the collective is the ability to interpret the biosemiotic dynamics that regulate the relationships between the actants.

2 The boundaries of the collective are defined by the biosemiosis itself.
3 The source of meaning is matter and its organic evolution.
4 This bio-collective is a black box (i.e., the dynamics of the collective are invisible).

Psychoneuroses are resistances, which can also be thought of as counter-programs of action that are opposed to those programs consolidated by the relations in the collective. For example, the child's drive is to love the mother. This drive has a defined program that involves a series of actions to be performed in order to love, or enter into symbiosis, with the mother. Psychoneurosis arises when the child's program of action is interrupted by a counter-program imposed by another object: the father. Resistance to the action program is thus generated. Psychoneurosis "explodes" if the child fails to overcome this resistance (i.e., he or she cannot find another program of action that can bypass the resistance). For example, the repressed drive can ally with material signs to strengthen itself and put pressure on resistance. The analyst must open up the collective by introducing elements, such as constructions and interpretations, which are new actants. These new actants try to form a broader alliance between drives and linguistic signs. More specifically, they are sets of linguistic and material signs that attempt to translate a set of counter-programs of action into linguistic signs and make it acceptable, in the sense that it can be accepted in the new conscious area of the collective.

Following Peirce's triadic model, I explain the emergence of the mind and its development from an organic stimulus. As explained in Chapter 1, the organic stimulus is the sign of a particular state of the body in its environment—in the sense that it refers to this state. According to Peirce, any sign requires and produces another sign (i.e., the interpretant), which mediates the relationship between the previous sign and its object. Therefore, the organic stimulus produces another sign: the drive as its interpretant. The drive is simultaneously a sign of the organic stimulus and its goal (i.e., the satisfaction). The concept of the interpretant introduces a potentially infinite reproduction mechanism. In fact, being a sign, the drive produces a new interpretant, the means by which it can reach its goal of satisfaction. The means here mediate between the drive and the satisfaction. This mechanism is potentially infinite; the drive always seeks a new means—interpretant.

The movement of the interpretants is a constant approximation of the goal (Figure 2.1). The evolution of the collective requires increasingly complex interpretants to define each other. More drives connected to each other form an ever-larger collective that requires translation processes, such as those seen earlier. There is no hermeneutic presupposition in this thesis. Meaning is a property emerging from some initial material conditions. From a biosemiotic point of view, the sign does not presuppose a meaning, as it is not a linguistic sign. Instead, it implies only the idea of a connection between things. Semiotics is a form of material

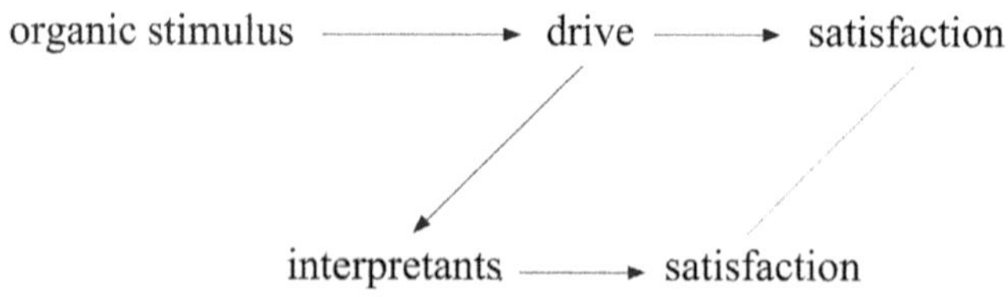

Figure 2.1 The basic form of a bio-collective.

engagement: "There are no meanings but only contexts, and more specifically networks of material engagement" (Malafouris 2013, 128).

I am not using concepts like "state of mind," "internal representation," or "intentionality." I am just describing a set of human and nonhuman actants that interact with each other and create alliances (strength = capacity to ally, according to Latour's scheme). The development of the mind leads to the definition of two areas in the collective: one that collects the set of action programs consolidated by the balance of power relations between the actants and another that collects the set of counter-programs of action and resistances that have not yet been overcome. Together, they continue to put pressure on the borders of the first area, thereby creating instability. The first area corresponds to the Ego and the second to the Id.

How can we explain the power relations and resistances in the mind? I use the four stages of translation identified by Callon (1986), which include Problematization, Interessement, Enrolment, and Mobilization. Problematization is the first stage, and this process is led by a focal actant who could be either an individual or a collective entity. The focal actant identifies the problem, its solution, and the relevant actants needed to solve it. It then creates an indispensable Obligatory Passage Point (OPP). The OPP defines the action program and the relationships that need to be established between the actants. The OPP also forms the basis for which the focal actor negotiates with other actants to conscript them into the actor network. The second stage is the Interessement phase. Here, the primary actant negotiates with the needed actant to get them to accept the roles assigned to them in the OPP. The third stage is the Enrolment stage. At this stage, the actants accept the roles assigned to them. If the Interessement stage is unsuccessful, the actor–network formation process either stalls or collapses. In this case, the actors do not accept the OPP; they propose other action programs or counter-programs. The last stage is the Mobilization stage, where representative actants emerge as spokespersons for black boxes in the actor network. The representative actant could be either an individual or more than one entity.

In the mind, the drive is the focal actant who acts to solve a problem, which is in this case its satisfaction. Psychoneurosis is nothing more than a deficit in translation or an interruption in negotiations, which can happen at any level.

I now translate Callon's four stages into the biosemiotics framework:

1 Problematization

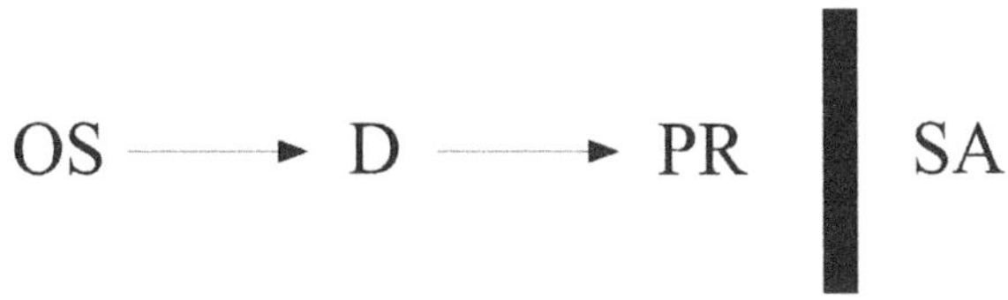

Figure 2.2 The organic stimulus (OS) produces a drive (D), which is its representative or its sign. However, there is a problem (PR) that prevents satisfaction (SA). The drive is the focal actant.

2 Interessement

Figure 2.3 The drive therefore seeks another form of satisfaction (SB) and to do so produces a new sign, an interpretant (int), that mediates the relationship between the drive and satisfaction.

3 Enrolment

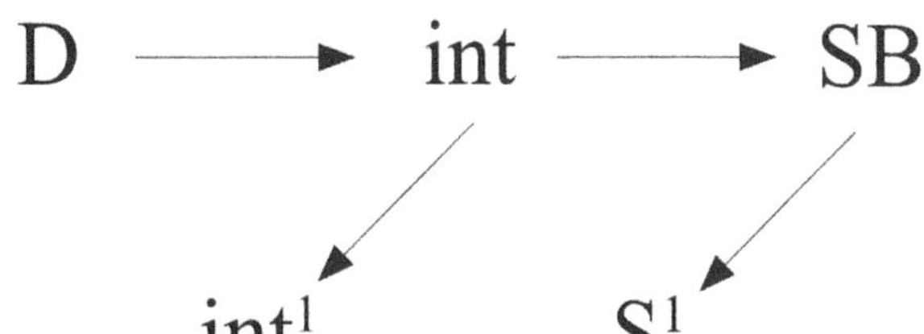

Figure 2.4 Following Peirce's principle of infinite semiosis, the interpretant produces a new interpretant and thus a new form of seeking satisfaction (S^1). The interpretant is a program of action. Through its materiality (e.g., affordances and shapes), it defines a way to reach the goal (SB) and, therefore, choose the following interpretants. The interpretant selects subsequent interpretants. Some interpreters who are candidates for the role are discarded; others are admitted. In this case, the sign is a material sign. According to Malafouris (2013, 97), "A material sign as an expressive sign does not refer to something existing separately from it but is a constitutive part of what it expresses and which otherwise cannot be known."

4 Mobilization

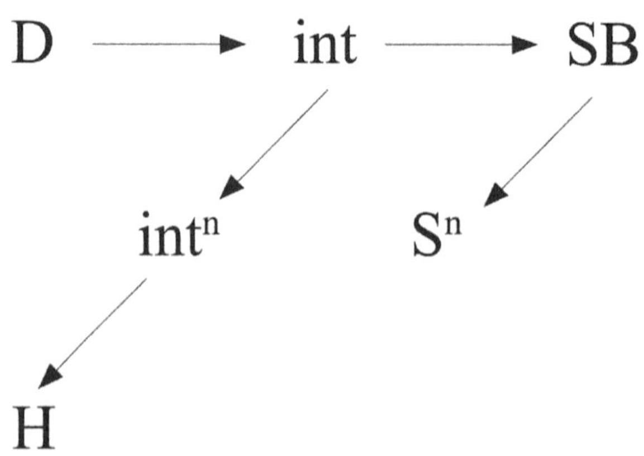

Figure 2.5 The process of succession of interpretants goes on until a point of stabilization is found, or until what Peirce calls a habit (H) is formed.

Nothing prevents the role of the focal actant from being assumed by a nonliving object, or an artifact, in the bio-collective. For instance, in *Playing and Reality*, Winnicott claims that in the world of children, there is no clear boundary between internal and external space or between subject and object. In the first phase of his–her life, the child experiences perfect unity with the mother and the surrounding environment. Baby and mother are a unity made up of two people. To be able to detach himself–herself from this fusional state and distinguish himself–herself from the rest of the world, and therefore establish his-her own identity, the child must "build a bridge" to the external world. The means by which this is achieved are neither wholly subjective nor wholly objective; rather, they are simultaneously objective and subjective. This is what Winnicott refers to as the "transitional object." The transitional object belongs to the external world (e.g., it could be a rag, blanket, word, lullaby, toy, and teddy bear); however, it is also the symbol of the mother's breast and the state of fusion with the mother. The child, according to Winnicott (2005, 45), needs this object to cope with reality and be able to endure the anguish resulting from the loss of contact with the mother. Thanks to the transitional object, the child accepts the loss of omnipotence enjoyed in the relationship with the mother. The artifact plays an active role; it makes it possible to redefine the drive and transform the action programs of the actants into the collective.

The MET reinforces this idea by emphasizing the relationship between the mind and material culture. Based on the work of Leroi-Gourhan

(1943, 1945, 1964) and Stiegler (1998), the MET overcomes the dualism between mind and body simply by erasing the boundary between these two entities. In contrast to the prevalent cognitivist, intracranialist, executive, and modernist definitions of mind, the MET claims that the mind and material things shape each other—this is material engagement. If the "mind evolves and exists in the relational domain as our most fundamental means of engaging with the world, then material culture is potentially co-extensive and consubstantial with mind" (Malafouris 2013, 77). This is not based on the assumption that there is a causal link between material culture, but that thinking is a compound made up of the brain and things. "Human thinking is, first and above all, thinking through, with, and about things, bodies, and others," remarks Malafouris (2013, 77). It is not only a causal logic that regulates the relationship between the mind and things but also a logic of emergence very similar to the one we found in Deacon: "thinking is not something that happens 'inside' brains, bodies, or things; rather, it emerges from contextualized processes that take place 'between' brains, bodies, and things" (Malafouris 2013, 77–78). Drawing upon ANT and Latour's work, the MET claims that minds and things are "constitutively interdependent—that is, that one cannot exist without the other" (Ibid.). In other words, those properties that we habitually attribute to "cognition" or "mind," such as feeling, memory, logical reasoning, or the ability to organize and select, cannot be explained if we look at the brain in isolation. Malafouris shows how material things and space have an influence, or agency, on the way the brain develops these properties. The MET can be seen as a philosophy of design; cognition arises from the interaction and mutual definition between brains, bodies, space, and material objects. Therefore, there are no isolated brains, bodies, spaces, or material objects, but only hybrids.

The criticism that can be directed at the MET is that of confusing interaction and participation. A clay tablet is an extension—an aid to cognition, and not cognition itself. If I use my smartphone to take a note, it is a support for my memory, not a constituent part of the memory itself. In other words, even if the smartphone has an active function and influences my way of remembering, this is not enough to be able to say that it is remembering—a constitutive part of remembering (Adams and Aizawa 2008, 2010). The MET instead challenges this idea; in this case, the causal or participatory link is based on an ontological link. The thesis according to which an object is only an "external amplifier" or "storage device" is wrong because "it leaves out the element that matters most: the extended reorganization of the cognitive system" (Malafouris 2013, 81). Objects not only help cognition but also transform and create it, so the smartphone not only helps my memory but also reconfigures it, shapes it, and introduces other previously absent operations into the process of remembering. Obviously, the ways the brain and a smartphone remember are completely different; however, memory arises from the interaction between them. This approach finds supporting evidence in current scholarship in palaeoanthropology, as in the work of Sterelny (2004) and Wrangham (2009).

Everything then becomes design in the sense of Latour (2008)—things are not created but are instead shaped and reshaped all the time. The brain works by continuously shaping and reshaping the space and objects around it. The latter, in turn, continuously shapes and reshapes the way the brain works. The mind, therefore, is not an object, nor does it have a location. Instead, it is the name we use to refer to a collective in constant movement. The mind derives from the collaboration between the body and matter:

> instead of seeing in the shaping of the handaxe the execution of a preconceived "internal" mental plan, we should see an "act of embodying." In tool making, most of the thinking happens where the hand meets the stone. There is little deliberate planning, but there is a great deal of approximation, anticipation, and guessing about how the material will behave. Sometimes the material collaborates; sometimes it resists. In time, out of this evolving tension comes precision and thus skillfulness. Knapping, then, should be seen more as an active "exploration" than as a passive "externalization" or "imposition of form." The knapping intention is essentially constituted through an act of collaboration between human and material agency, one of the earliest manifestations of human "tectonoetic awareness."
>
> (Malafouris 2013, 235–236)

I believe that Malafouris's (2013) theory is very close to Stiegler's conception of epiphylogenesis. Therefore, I propose to connect biosemiosis, MET, and epiphylogenesis as three levels of the same evolutionary scheme (Figure 2.6). Technology arises from the encounter between natural semiosis and material engagement, but it is more than that. The artifact is a materialization of memory (bodily, cognitive, affective, etc.), which in turn transforms the previous two levels. Epiphylogenesis introduces a historical and transformative dimension in the relationship between natural semiosis and material engagement. For this reason, epiphylogenesis also puts forward a stratification of ever-deeper levels of natural semiosis and material engagement. Some of these levels are still accessible to consciousness, but others are not. Complex artifacts have many layers of different epiphylogenesis, with certain ones being conscious and others unconscious. It is the

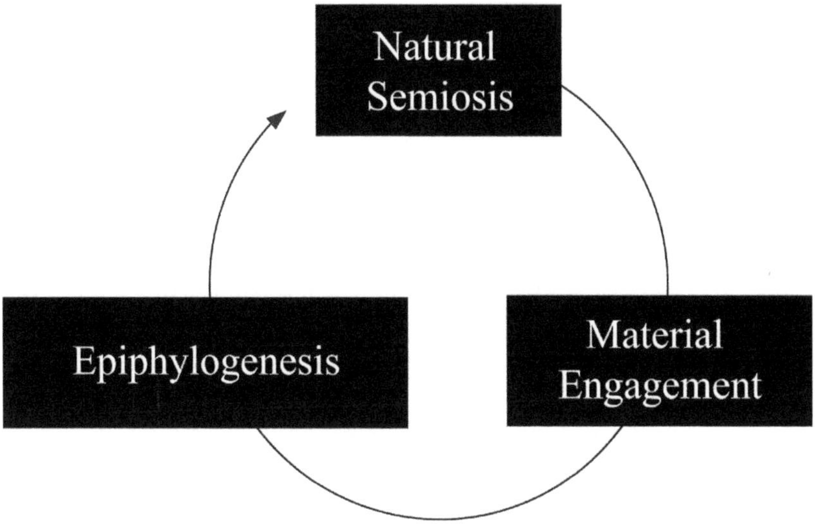

Figure 2.6 The three levels of the constitution of the mind: the natural conditions of semiosis, the interaction with matter, and the historical evolution of technology.

task of the archeology of the mind (Malafouris 2010) or of psychoanalysis as technoanalysis to explore the unconscious dimension of epiphylogenesis.

Did Freud think in this way? The importance of *Beyond the Pleasure Principle* in the development of psychoanalytic thought is well known. In this essay, Freud introduces a crucial concept: the compulsion to repeat—that more original, more elementary, and more instinctive force than the pleasure principle and its modification of the reality principle. The compulsion to repeat is

> an ungovernable process originating in the unconscious. As a result of its action, the subject deliberately places himself in stressing situations, thereby repeating an old experience, but he does not recall this prototype; on the contrary, he has the strong impression that the situation is fully determined by the circumstances of the moment.
>
> (Laplanche and Pontalis 1988, 78)

In introducing the concept of a compulsion to repeat, Freud uses the example of his grandson and his way of playing. In fact, the child had a habit of throwing away all the small objects he could get hold of into a corner of the room. In doing this, he emitted the sound "o-o-o," which was accompanied by an expression of interest and satisfaction. The child used all his toys just to throw them away—that was the game. Immediately after this general description, Freud introduced an observation. I report the complete passage here because it is essential for my thesis:

> One day, I made an observation that confirmed my interpretation. The child had a wooden reel, with some string tied around it. It never crossed his mind to drag it along the floor behind him, for instance, in other words, to play toy cars with it; instead, keeping hold of the string, he very skillfully threw the reel over the edge of his curtained cot so that it disappeared inside, all the while making his expressive "o-o-o-o" sound, then used the string to pull the reel out of the cot again, but this time greeting its reappearance with a joyful Da! ("Here!"). That, then, was the entire game—disappearing and coming back—only the first act of which one normally got to see; and this first act was tirelessly repeated on its own, even though the greater pleasure undoubtedly attached to the second.
>
> (Freud 2003, 88)

The artifact, or the reel, introduces a change in the development of the drive. Through the mediation of the object, the satisfaction of the drive is postponed, and renunciation is made possible. The affordances of the object shape the drive (to allow the mother to leave) and endure renunciation. However, there is another point to consider: the affordances also allow us to go "beyond the pleasure principle" and give voice to an even deeper drive, which is the compulsion to repeat. I do not want to analyze the content of Freudian theories or evaluate their reliability here. What interests me is Freud's method. The child's behavior is considered a sign of the drive (i.e., the compulsion to repeat). This drive is in turn the sign of a deeper biological reality, that is, a general conservative tendency—according to Freud—of organic life. The artifact is not an element external to this drive dynamic; instead, it is an agent that contributes to the development of the drive and the achievement of its goal of satisfaction. When the drive, the evolution of life, finds an obstacle, it deviates and uses objects to do so. This deviation has a semiotic character, as the reel is the interpreter of the drive; it is what still allows the drive to move toward its dynamic object.

In *Beyond the Pleasure Principle*, the reel has a sense of agency because it expresses the resistance of the drive to the obstacle that prevents its immediate satisfaction. If agency is the measure of resistance, passivity is the lowest level of agency in not resisting. Strength would make no sense if there was no resistance. Therefore, causality and intentionality are not the prerequisites of agency, but the consequences of it; in other words, they are forms of translation. The drive deviates and transforms itself; it is the focal agent who guides the network formation process. The same thing happens in the clinical case of the wolfman. In the dream at the center of the analysis, the only movement is that of the window that opens by itself; the wolves remain motionless. The window acts and shows the terrifying image of the silent wolves. The human subject is an actant, among others, that follows the biosemiotic process.

Conclusions

The goal of this chapter was to extend and transform the results of Chapter 1. Psychoanalysis now appears to be very different from what we are used to. The Freudian methodology was the starting point for a complete redefinition of the problem of mind/brain, mind/body, and matter/meaning relations. Freud himself directed us toward Latour (i.e., toward a network theory). The mind is the name of a collective made up of humans and nonhumans, in which meaning and agency are distributed as emerging qualities among the different actants of the network. "The human mind is a product of biological evolution as much as it is an artifact of our own making" (Malafouris 2013, 231). As we will see in the next chapter, technoanalysis consists of studying technology as a fundamental actant in the collective called mind. This approach intends to recover the essence of the Freudian method of exploring the collective mind. As we will see in the next chapter, the ontological turn introduced by ANT allows us to recover and expand the use of the Freudian method and its logic. Moreover, the hybridization introduced by ANT also allows us to conceive of the problem of suggestion in a new way (i.e., by considering it as something distributed in the network), since suggestion is also induced by nonhuman actants.

I summarize the main findings of this chapter as follows:

- Psychoanalysis is a method for studying a certain type of bio-collective.
- I define technoanalysis as the study of the bio-collective mind when the focal actant is a nonliving and nonhuman entity that is capable of (a) respecting the conditions of admissibility in the bio-collective and (b) producing agency and meaning autonomously. In even more precise terms, technoanalysis identifies and examines the resistances between action and counter-action programs in the mind. In doing so, technoanalysis uses the Freudian experimental method and its logic. In the next chapter, I will show the application of technoanalysis in the field of social robotics, with the particular objective of examining the problem of interaction failure in HRIs.

3 Mediation and Anti-Mediation

Google Glass, the Metaverse, and Social Robotics

This chapter constitutes the central part of the book, in which I move from the definition of the theoretical model of technoanalysis to its application. In this chapter, I will (1) define a descriptive method in line with the theoretical assumptions previously established and (2) apply this method to three case studies: Google Glass, the Metaverse, and social robotics. In the next chapter, I will apply the descriptive model to the study of an AI-based chatbot. My goal is to show not only the originality of technoanalysis but also its connection to interactive design.

In the previous chapter (see Table 3.1), I outlined a philosophy of technology based on ANT and biosemiotics. I first introduced the concept of the bio-collective through a critique of the ANT of Latour. Using the MET, I connected the concept of the mind to that of the bio-collective—the mind is a bio-collective. Drawing up Stiegler, Canguilhem, and Simondon, I examined the tendency of the bio-collective to engage in artificialization, that is, to extend biological codes and information through networks of non-living actants. The mind is not a compact substance closed in the skull of the human being; it is instead pixelated, reduced to a set of non-linear interactions between humans and nonhumans endowed with agency, intentionality, and the ability to produce signs.

Psychoanalysis has changed its face. It now appears to us as the exploration of a certain part of the bio-collective called the mind, that place where biological information resides that is not translated into human symbolic form, which is the language that we know (see Chapter 1, the analysis of Deacon's analysis). This information is an immense set of signs produced by biological and material conditions, and yet these signs do not translate into human language or, therefore, into preconscious and conscious terms, even if they influence them. The Freudian unconscious is a small part of this set. Following the results of Chapters 1 and 2, i.e., our reinterpretation of psychoanalysis through biosemiotics and ANT, we must admit that the notion of the unconscious—the object of psychoanalysis—has radically transformed. The unconscious is now (a) that part of the bio-collective that does not translate into human symbolic

DOI: 10.4324/9781003345572-5

Table 3.1 The fundamental concepts of the book, which compose the theoretical framework through which we reinterpret psychoanalysis.

ANT	*Semiotics*	*Biosemiotics*	*Technology*	*Theses*
Symmetry Critique of Latour's semiotics Broadening ANT through MET	Peirce Interpretant Infinite semiosis Absence	Semiosis as a material process Biological information Umwelt Logic of emergence	The artifact as a semiotic structure The technical tendency Epiphylogenesis	Any artifact is part of the umwelt Any artifact is based on semiotic structures Any artifact transforms the umwelt

structures and (b) that part of the epiphylogenetic stratification of the artifact that is not accessible to consciousness.

The objection that the thesis, "Any artifact is part of the umwelt," is contradictory because it is anthropocentric is wrong. In fact, as I explained in the previous chapter, the umwelt is a co-construction of the actor and its environment. The human umwelt is composed of semiotic structures that are not only human; for instance, it contains semiotic networks of plants, animals, cells, and material objects. Furthermore, the umwelt is a model, an invention that is constantly updating and undergoing transformation.

Biosemiotics allows us to (a) overcome the rigid dualism of matter/meaning and (b) admit the idea of biological information, that is, of a type of information that is not necessarily linguistic, symbolic, human. It would obviously be too complex to develop a detailed analysis of the literature and debate on the concept of biological information as a whole (Godfrey-Smith and Sterelny 2016), and doing so would go beyond the limits of this research. As such, I limit myself to the analysis that Barbieri gives (2012, 2015). Barbieri challenges the chemistry paradigm—that is, the idea that life is only chemistry, as opposed to the idea that life is chemistry plus information plus codes. According to Barbieri, information is a cellular artifact; it is produced by cells from a code, the genetic code. Biological information is a sequence of genes and proteins defined by a set of correspondence rules, according to certain chemical conditions. This information is observable, but it is neither measurable nor computable. Also, this kind of information is not something that can be derived from simpler forms: "Genes and proteins are never produced by spontaneous processes in living systems. They are produced by molecular machines that physically stick their subunits together according to sequences and codes. They are manufactured molecules, i.e., molecular artifacts" (Barbieri 2012, 148). The crucial point is that "the manufacture of genes and proteins requires sequences and a genetic code," and "we cannot describe living systems without sequences and coding rules" (148). Starting from this idea, Barbieri (2015) defines a series of increasingly complex organic codes that produce increasingly complex information and meanings: "The existence of many organic codes in Nature is therefore an experimental fact—let us never forget this—but also more than that. It is one of those facts that have extraordinary implications" (xiv). The existence of organic codes is the result of what Barbieri calls *codepoiesis*.

What is the relationship between biological information and technology? The thesis of this chapter is that, if we admit the assumption common to post-phenomenology and ANT, namely, that "technological mediation is part of the human condition—we cannot be human without technologies" (Verbeek 2015, 3), then we must admit that technological mediation is a process of the constant translation and re-translation of biological information. This means that identifying the fundamental forms of technological mediation would make us able to identify and analyze the forms of biological information that flow in them. Therefore, in the first part of this chapter, I will analyze the theory of mediation starting from post-phenomenology. I will show that the post-phenomenological concept of mediation is compatible with ANT in the sense that, beyond a radical diversity on the ontological level, the post-phenomenological theory of mediation can be used in ANT on the methodological level to recognize and describe the different kinds of association and translation present in the bio-collective.

My argument is as follows: *if* we admit that the fundamental types of mediation described by post-phenomenology identify the main forms of the artificialization of life—the "technical tendency," following Stiegler's terminology—*then* studying such mediation will give us some cues to identify and study the dynamics of the biological information that circulates in the bio-collective and, especially, what I will call the forms of *anti-mediation*, or the biological information that does not translate into mediation. Anti-mediation is not a malfunction. However, the malfunction may be an indication of an anti-mediation. My claim is that anti-mediation is an unconscious phenomenon, that is, a translation error in the collective, in the network, and in the relationship between human and machine. Technoanalysis views this error as the manifestation of an epiphylogenetic stratification of the artifact that is not accessible to consciousness. Forms of anti-mediation are the object of technoanalysis.

The immediate objection is that this argument betrays the symmetry of ANT and anthropomorphizes our research. However, I do not think that this objection is valid. The resistance to anthropomorphization depends on the way we understand mediation, that is, on the direction in which we read it. ANT pushes us to broaden the post-phenomenological concept of mediation and, therefore, to read every form of mediation in the opposite sense—that is, no longer from humans to machines but from machines to humans.

Mediation and Networks

The concept of mediation is crucial for post-phenomenology. "Post-phenomenology builds on classical phenomenology and American pragmatist philosophy to articulate deep descriptions of the various ways that human experience is transformed through technology usage" (Rosenberg 2018, 174). The central thesis of Ihde (1979, 1990), Verbeek (2015, 2011), Borgmann (1984), and Rosenberg (2018) is that human contact with reality is always mediated and that technologies offer a form of fundamental mediation. Technologies are not mere tools but actively define human practices and experiences: "Designing technology is designing human beings: robots, vacuum cleaners, smart watches—any technology creates specific relations between its users and their world, resulting in specific experiences and practices" (Verbeek 2015, 23). Therefore, technologies "are conceived as transformative mediators of experience, coming between the user and the world, and changing the possibilities for perception and action" (Rosenberg 2018, 174). In doing so, post-phenomenology reconceptualizes the Husserlian notion of intentionality by radicalizing it; the relationship itself is the source of the polarity between subject and object—not the other way around (Rosenberg and Verbeek 2015, 12). With this operation, however, post-phenomenology recovers and betrays Husserl at the same time. It recovers the Husserlian notion of intentionality because it updates its relational vision; however, this theoretical decision betrays Husserl because inserting mediation into the heart of intentionality implicitly means losing the characteristics of immediacy and realism that constitute the essence of intentionality in Husserl—at least in the *Logical Investigations* (Husserl 2012).

Starting from these premises, post-phenomenology aligns with ANT in its critique of the modern split between subject and object in which they are understood as two separate poles. However, the post-phenomenological approach does not pursue the symmetry between humans and nonhumans, as does ANT. The distinction remains, despite mediation and mutual constitution, and "When we give up this distinction, we also give up the phenomenological possibility to articulate (technologically mediated) experiences from within" (Rosenberger and Verbeek 2015, 20). While ANT studies the relationships between humans and technology in the third person, post-phenomenology studies this relationship in the first person, from the point of view of the mediated human subject, who remains the source of meaning. Technology mediates the *human* relationship with the world.

Table 3.2 The types of mediation that technology (T) operates in the relationship between humans (H) and the world (W).

Relationship	*Structure*
Embodiment	(H-T) → W
Hermeneutic	H → (T-W)
Alterity	H → T(W)
Background	H (T/W)
Cyborg	(H/T) → W
Immersion	H ←→ (T/W)

An important aspect of the phenomenological method is the study and classification of the different forms of relationships between human beings and technology (Ihde 1990; Verbeek 2015). Table 3.2 summarizes the basic types of mediation and indicates their formalization.

In the embodiment relationship, technology is directly connected to the body and, together with the body, establishes the relationship of humans with the world; glasses are a typical example. In the hermeneutic relationship, humans must read and interpret the signals of technologies to access the knowledge of the world (e.g., the use of magnetic resonance imaging). In the relationship called alterity, or otherness, human subjects interact only with technology, while the world remains in the background, separate (e.g., HRIs). In the relationship called *background*, technology constitutes the general context in which humans act; there is no direct interaction between humans and technology, but technology entirely defines the humans/world relationship (e.g., air conditioners or windows). The cyborg relationship is one in which the relationship between technology and humans is even deeper and closer, so that it becomes a symbiosis (e.g., implants in the brain). In contrast, when technology is not just in the background but is an agent capable of interacting with human action, there is an immersion relationship (e.g., a smart house, which is not in the background, as is simple air conditioning); technology defines the world and, together, they interact with the human.

Beyond the ontological differences, post-phenomenology and ANT can be partially integrated. "Actor-network theory is primarily interested in unraveling the networks of relations by virtue of which entities emerge into presence, while a postphenomenological approach, by contrast, seeks to understand the relations that humans have with those entities" (Verbeek 2005, 164). The taxonomy of mediation types can provide a useful tool for analyzing the types of associations in networks. I mean that the notion of mediation and that of a network can be integrated. Postphenomenologists indeed affirm that "technologies always remain 'multistable,' that is, they have the capacity to mediate human relations to the world in multiple ways" (Rosenberger 2018, 175). The meaning of technological mediation depends on the stabilization of the power relationships in the network of actants and, therefore, on the negotiation/translation processes. As Cathrine Hasse explains, "in postphenomenology, multistability refers to technologies that vary in how their meanings are stabilized as they cross time. . . . Multistability emerges in the meeting of different kinds of practices in relation to the affordances offered by the design" (2015, 164).

This does not mean betraying the principle of symmetry. A betrayal of the principle of symmetry would be to consider the human being as the only source of meaning in that relationship—that is, the only entity with agency, autonomy, and the ability to produce meanings. In other words, I argue that, while maintaining a symmetrical ontology, we can use some conceptual tools of post-phenomenology as hermeneutic resources to understand the forms of associations in the bio-collective.

This does not mean either that the postphenomenological notion of multistability denotes the fact that technologies, as concrete material objects, enable just any mediated experience; "a technology cannot be used for just any purpose, be meaningful in just any manner, or facilitate just any perception. The concrete materiality of the device places limitations on what kinds of relations can occur" (Rosenberger 2018, 175). I claim that the association and translation process in the network defines the possible stability of a specific technology in that network. Identifying a mediation means identifying a certain stability in the relationship between an artifact and its context.

When we move from a single interaction to complex interactions, we must abandon the phenomenological point of view and assume that of ANT. When we consider vast sociotechnical systems, we are dealing with a complex network of human and nonhuman actants who have many different relationships. Hermeneutic, embodiment, background, alterity, cyborg, and immersion relationships intertwine and condition each other. This also means that even nonhuman actants can develop autonomous agency—that is, the ability to influence these mediations and reverse their meaning. We therefore need a precise descriptive method referring to a defined context. From Latour's point of view, mediation is translation, so the forms of mediation we have described are also forms of translation. Mediation succeeds when it can translate biological information into another semiotic system—for example, the characteristics of the artifact itself, as in the case of Winnicott's transitional object. Mediation fails when this translation fails. Furthermore, following Latour's terminology, we must also admit forms of anti-mediation, or translation errors that prevent the flow of biological information in the bio-collective. In this case, the mapping between the two sets of properties (i.e., the drive and the artifact) fails. Applying the Freudian technique, technoanalysis intends to show the connection between forms of anti-mediation and the unconscious.

In the next sections of this chapter, I intend to show that this connection between the post-phenomenological theory of mediation and ANT can be used as a toolbox, in two senses:

1. A descriptive sense: An artifact can be described as a concatenation of mediations—mediation is the basic element of the artifact. The different actants in the bio-collective establish different types of mediation and anti-mediation.
2. A therapeutic sense: A mediation in an artifact can be replaced by another mediation or by a concatenation of mediations; we can transform the artifact by replacing or transforming the mediations present in it. An anti-mediation can be "cured" by replacing or transforming mediations already present in the artifact.

Technoanalysis is therefore both a form of reverse engineering (from the artifact to its description, to identify the fundamental mediations) and a form of design that continuously redesigns the artifact by transforming or replacing the mediations already present.

Google Glass and the Metaverse: The Technological Uncanny and Digital Wilderness

Why did Google retire Google Glass in 2015? The first version of Google Glass was a failure, in the sense that it did not get the expected response: "It was a bold attempt to bring the world a step further into the information age. The idea was great, but the execution and development weren't" (Doyle 2016, 2). As evidenced by Google's withdrawal of all Glass-related media on social networks, "it was clear that the revolutionary product wasn't performing as planned" (1). Google Glass has two basic functions: "to quickly capture images and to have a feed of useful information from the internet a glance away. What are the most practical daily uses for these features? None" (1). The failure of Google Glass "is due to the lack of clarity on why this product exists. The designers did not clearly define or validate: the users' problems, what solutions Google Glass would provide for its users, or how customers would use the glasses" (Yoon 2018). We can summarize the problems encountered as follows:

1. Health concerns about the risk of harmful radiation
2. Security and privacy concerns
3. Lack of clarity on the scope of the artifact (i.e., what is it really good for?)
4. Battery limit (only four hours of life)
5. High price

6 Language problems (the device only works for native English speakers from the US or UK, meaning that it does not receive commands in other languages or versions of English)
7 Temperature problems (the device overheats)

Google Glass is a technology unlike any other; it is very invasive, and its intent is not only to propose a new way to access augmented reality but also to put this access in close contact with the human body. Google Glass intends to merge technology and eyesight, transforming human vision. This action program—using the terminology of Latour and Akrich (1992)—involves many sub-programs (i.e., audio commands, video camera, etc.). The device failed because the counter-programs took over. These counter-programs resulted only in part from nonhuman agents (e.g., the lack of battery life). The strongest counter-programs came from humans and, precisely, from the relationship between Google Glass and the human body. An analysis of the comments on promotional videos for Google Glass on YouTube confirms this thesis. In fact, there are two, often mixed, trends in these comments: enthusiasm about the novelty of the technology and the fear of damage to the users' eyesight, brain, and body in general—a general sense of uncanny.[1]

If we take a Latourian perspective, Google Glass comprises a chain of human and nonhuman agents in constant negotiation. The drive—in this case, the sense of insecurity and fear caused by the glasses in humans—plays a key role; it is an actor that "blows up" the device, in the sense that it triggers numerous failures, such as disappointed users (Akrich and Latour 1992, 259). The drive forced Google's engineers and designers to redefine the project and relaunch it years after the first version. Google Glass is an example of a failed bio-collective. The biological information, the umwelt, was not translated by the other actors—for example, in the choices made by the designers. Therefore, the device did not stabilize. The lack of integration between technology and the human body occurred not because of a technical problem but because of a semiotic problem. Users not only failed to understand the main function of Google Glass but also feared its effects on their health and privacy. Fear and a sense of uncanny were the breaking point.

Let us now try to translate this analysis into terms of the mediation theory defined earlier. It is not just the embodiment relationship that played a key role in the case of Google Glass. The situation is much more complex. To the embodiment relationship are added the hermeneutic relationship and the cyborg relationship:

$$[(H\text{-}T) \rightarrow W] + [H \rightarrow (T\text{-}W)] + [(H/T) \rightarrow W]$$

In the case of Google Glass, the triple mediation acts as a movement of belonging and distancing between humans and technology. In the embodiment mediation, there is a fusion between the human body and technology, while in the hermeneutic mediation, "technologies form a unity with the world, rather than with the human being using them" (Verbeek 2015, 29)—therefore, we are witnessing a rupture of unity between the human being and the artifact. The third relationship synthesizes the previous two, translating them onto a completely new level. The cyborg mediation is not merely a fusion of technology and living organism but a complete redefinition of both that abolishes any boundary between them. Therefore, the cyborg mediation goes beyond embodiment and hermeneutics and intends to redefine the entire human identity: "Rather than being a technologically mediated form of human intentionality, it is the intentionality of a new, hybrid entity: a cyborg" (Rosenberger and Verbeek 2015, 21).

We must then reformulate the Google Glass structure as follows:

$$[(H/T) \rightarrow W] + \{[(H\text{-}T) \rightarrow W] + [H \rightarrow (T\text{-}W)]\}$$

Google Glass transforms not only our context—that is, our relationship with things in our surroundings and how we "see" spaces in front of us—but also our own subjective identity. To accomplish this goal, the embodiment relationship and the hermeneutic relationship were used to build a new cyborg-like relationship. What triggered the fear reaction in the users? The fact that this synthesis failed. The increase in the number of technological mediations resulted in what I call an *anti-mediation*—that is, an opposition to the technological mediation, or a short circuit in the mediation. There is a threshold of acceptability beyond which technology becomes something incompatible with the human being—a sort of uncanny valley. To respond to this crisis of acceptability, it is necessary not to multiply mediations but to know how to balance them.

The origin of anti-mediation lies in the cyborg relationship, which is the true dominant form of mediation. Cyborg mediation fails to synthesize embodiment and hermeneutics. In the case of Google Glass, the accumulation of different mediations without a synthesis created a sense of over-investment in the consumer, with the consequent increase in tension. To decrease the tension, it is necessary to act on the basic mediations (i.e., embodiment and hermeneutics) to improve the perceptual dimension of the glasses or the way in which to "read" the glasses and the information they convey. Acting on the basic mediations, transforming them, allows the synthesis to be improved.

For example, a solution could be to introduce a new mediation, such as a background mediation, by connecting the glasses to other devices or services, creating a more integrated technological umwelt. Integration will obviously depend on the psychic response. The design must therefore be capable of acting in three different directions: visceral, behavioral, and reflective (Norman 2004). Another solution could be to introduce a hermeneutic mediation of another type, such as a narrative that better configures the identity of Google Glass (i.e., the history of its design and creation, etc.), making it more acceptable to users. Another solution could be to transform the embodiment mediation by using different materials in the glasses themselves.

As can be seen from this example, anti-mediation can have many forms such as the clash between mediations or the failure of synthesis of mediations, which does not necessarily imply a clash.

The failure of synthesis can be explained in many different ways. For instance, the evolutionary esthetics hypothesis claims that humans are highly sensitive to visual esthetics; this hypothesis suggests that selection pressures have shaped the human preference for certain physical appearances signaling fitness, fertility, and health (Ferrey et al. 2015). The pathogen avoidance hypothesis (MacDorman and Ishiguro 2006; MacDorman et al. 2009) may also be pertinent, according to which troubles with technology must be related to our instinct for self-preservation. According to this view, for instance, visual anomalies in human replica droids—which are perceived as being genetically very close to humans—provoke disgust because an evolved mechanism for pathogen avoidance detects these anomalies as indicative of risk or danger.

These hypotheses do not seem convincing to me because they only consider isolated aspects of the problem. The concept of the technological uncanny (Aydin 2021) seems more compelling. Reinterpreting the concept of the uncanny valley (Mori et al. 2012), Aydin claims that the uncanny is "an alterity within that cannot be simply explained in terms of something external that challenges or influences our internal convictions, preferences, values or goals" (Aydin 2021, 206). Therefore, the uncanny "cannot simply be opposed to the canny: *heimlich* and *unheimlich* are not simply opposites, since *unheimlich* signifies the concealed and the hidden and, at the same time, the familiar and domestic" (206). I argue that the fear and sense of uncanny caused by Google Glass are symptoms of this more general phenomenon: the technological uncanny.

Starting from Lacan (2005), and Nancy (2010), Aydin argues that the uncanny is an existential dimension that defines the human identity; it is a profound, radical, unbridgeable otherness that is a condition for the formation of the self: "we do not completely possess the self that we attempt to form" (Aydin 2021, 191). It is the paradox highlighted by Lacan's concept of the "mirror stage"; the other is at the same time a necessary condition for the formation of the self and an obstacle to achieving a complete unity of the self. The self is therefore originally contaminated, externalized. The uncanny feeling experienced in the relationship with a robot too similar to a human being is therefore motivated by the fact that that extreme similarity evokes the constitutive otherness of the human identity—that is, it reveals the fragility of the human identity, how small is the difference between humans and nonhumans.

According to Aydin, this is an aspect that characterizes all forms of technology. The artifact, in its being at the same time subject and object, evokes something repressed and unacceptable: "In confrontation with the humanlike robot, I not only become aware of what makes me different from it, but also of the impossibility for me to appropriate that difference" (Aydin 2021, 208). In the confrontation with the strange and, at the same time, familiar robot, "the self not only uncannily senses the human in the robot, but also *the robot in the human*" (208). Technology redefines the human identity by fragmenting it, disorienting it, and, thus, causing a sense of total disorientation, which is the return of a much more original ontological condition. Google Glass represents the appearance of an incomprehensible actant, whose program of action is not completely defined and which, at the same time, arouses a feeling of inadequacy and rejection in users.

The case of Google Glass also leads us to consider another decisive aspect. The technoanalyst must ask themselves not only "What kind of mediation does technology create between the human being and the world?" but also "What kind of mediation does the human being realize exists between the machine and the world?" and "If technological mediation for the human being is hermeneutic, what kind will human mediation be for technology?" This also implies not only that technology must adapt to the human but also that the human must adapt to technology. Further, it implies that anti-mediation may come from the technology and not just from the human. Therefore, it is necessary to improve both the technological mediation for the human and the human mediation for technology.

In Table 3.3, I rework the forms of mediation that I previously illustrated (see Table 3.2), in the opposite direction. The techno-centric hermeneutic relationship, for example, is that of the AI systems that define our tastes and purchases on the Internet; technology defines the world and, therefore, shapes the human being according to this world. The techno-centric relationship of alterity is even more radical, as technology refers to the world completely excluding the human being. A perfect example of this situation is the set of relationships between networked machine learning algorithms that exclude the human decisions made in the New York Stock Exchange. The techno-centric relationships of embodiment, cyborg, and immersion have similar structures to those in the first scheme; only the power relations differ. A smart house, for example, shifts from being H ⟷ (T / W) to being T⟷ (H/W) when H transforms its habits, lifestyle, and thinking according to those imposed by T.

To further illustrate this idea, a social media platform such as the Metaverse has the following structure: T(W) → H. The Metaverse is a form of technology that fully defines the conditions of the world, and

Table 3.3 Techno-centric forms of mediation.

t-*Relationship*	t-*Structure*
t-Embodiment	(T-H) → W
t-Hermeneutics	(T-W) → H
t-Alterity	T(W) → H
t-Background	T (H/W)
t-Cyborg	(T/H) → W
t-Immersion	T ←→ (H/W)

humans must adapt to those conditions. The Metaverse is both the network and the focal actor of the network at the same time. In such a case, the network becomes an actant and shapes the other actants. The Metaverse is not a tool but an agent that redescribes and reconfigures the personal and collective umwelt. "The Metaverse will be the successor to the mobile Internet. We will be able to feel present—as if we are right there with people, no matter how far apart we are. We will be able to express ourselves in joyful and completely immersive new ways," said Mark Zuckerberg, the co-founder and chief executive officer of Meta Platforms.

> Today we are seen as a social media company, but in our DNA we are a company that builds technology to connect people and the Metaverse is the next frontier just like social networking was when we started. Our hope is that within the next decade the Metaverse will reach a billion people, host hundreds of billions of dollars of eCommerce and support jobs for millions of creators and developers.[2]

The case of the Metaverse is not isolated; video games such as Fortnite and The Sandbox are already parallel universes in expansion, wherein the same relationship, T(W) → H, is always applied. With the Metaverse, we have overcome the concept of cyberspace. Nusselder (2009) distinguished the *matrix* as the "noumenal" dimension of codified objects consisting of zeros and ones (i.e., the database); *cyberspace* as the "phenomenal" mental space of the conceptualization or representation of code objects; and the *interface* as their crucial medium. In the Metaverse, the matrix has folded in on itself in such a way as to redefine all three of these aspects. Cyberspace—that is, the human dimension of the matrix—becomes an ever-changing interface within the matrix, that is, between the matrix and itself.

The Metaverse is not another example of virtual reality. It does not intend only to question the differences between the virtual and the real or the identity of the self (e.g., the dissemination of the self in the multiple user domain). The Metaverse has another goal: completely redefining the world and experience. It builds another world in which human beings are hosted, accepted, and welcomed. This is a vitally important point; the Metaverse is (or is proposed to be) an entirely autonomous, uncontaminated digital world that humans are called to explore and adapt. Zizek (2008, 170–172) claims that the computerization of experience transforms three "lines of separation" that constitute the hermeneutic horizon of experience itself: (a) the separation between natural and artificial life; (b) that between objective, true reality, and its simulation; and (c) that between accidental feelings and affects and the core of subjective identity. I think that the metaverse redefines all of these lines on another level. According to Zizek (2008, 172–173), the digital puts into question the essential distinction between inside and outside, between the subject and its outside; this is an example of castration. Zizek refers to the Lacanian theory of the phallus as a signifier of desire. The Metaverse takes this digital trend to its extreme consequences; it is a new sphere, to use the terminology of Sloterdijk (2011–2016), which encompasses both the external and internal, so that they become only effects of the sphere. This sphere is no longer a "second nature," according to the well-known Marxist formula (Zizek 2008, 173), but even a "third nature."

I argue that this third nature is connected to a contemporary myth, that of the "digital wilderness." Zizek (2008, 167) argues that there are two contemporary myths connected to the rise of digital technology: (1) through cyberspace, we are returning to wild thoughts, *pensée sauvage*, based on a montage of different perceptual fragments ("As not hypertext a new practice of montage?" Zizek (2008, 167) asks) and (2) through cyberspace, we are moving from the modernist culture of computation to the postmodern culture of simulation and dissemination, that is, a network without a center, where processes are always complex and non-transparent. These two myths, great constructions of the imagination, are present in the Metaverse, but in a minor tone. The real dominant myth in the Metaverse is the idea of an autonomous, pure digital world, uncontaminated by humans' presence in which humans are only guests. This is the dream of a new frontier where everything is possible, but this frontier is built by the machine, not by humans.

The elision, or removal, of humans from the digital is an important feature of contemporary digital imagery. This is what Taylor (2019) highlighted when introducing the concept of the "technological wilderness." He studies the stock images of data centers disseminated in the mass media. What is striking is the dystopian representation of these structures, which are portrayed as new pristine lands, IT jungles totally automated and lit only by neon. "Images of data centers persistently focus on their futuristic furnishings and the high-tech IT equipment they contain, rather than the people that work in these buildings" (Taylor 2019, 2). Another noteworthy element is the removal of human beings. Humans almost never appear in these images. The workers who are needed to maintain these facilities have disappeared. In this way, stock images of data centers do not represent reality; instead, they repress it. These images represent an idea of the future characterized by the disappearance of humans and the domain of technology. "Visual images produced and released by data centers are thus valuable ethnographic objects that provide an insight into how the industry narrates itself and attempts to govern the conditions of imaginative possibility through visual media" (Taylor 2019, 5–6).

The presence of humans is marginalized because they are considered unreliable; most of the security problems in data centers are caused by human error. Thus, humans are considered obsolete; for the sake of the data center, humans must be excluded or even eliminated.

> Amidst these technological futurescapes, traces of human life are rarely to be found. The spaces in the data center associated with human activity, such as the offices, waiting areas, water closets, meeting rooms, kitchenettes, and cleaning cupboards, are persistently absent from these image collections.
>
> (Taylor 2019, 8)

As Buchli (1999) taught in the case of socialism, material culture has profound symbolic and ideological meanings. The images of data centers convey a visual strategy, a symbolic economy that promises an entirely nonhuman world where the responsibility of everything is owned by machines. In the images of empty data centers, contemporary humanity projects an illusion; we externalize an intense desire for self-destruction, a death drive. These images help us to bear this urge, to make it acceptable by hiding it. The human subject is no longer just decentralized, but eliminated altogether—or accepted as a marginal aspect. The phantasmatic scenario of the "digital wilderness" is that the Earth is no longer the place, or the sphere, of the human being. The human being is no longer comfortable here. For this reason, humans must look to other planets, other spaces, to an increasingly complex dimension elsewhere. In other words, in order to survive—to allow human social life—humans must surrender control to machines.

What Is Anti-Mediation? How Can We Recognize It?

I now intend to give a more precise definition of the concept of *anti-mediation*. Let us say that if a certain technological mediation prevents the actor network from being held together, and it is not a solvable technical error, then that mediation is an anti-mediation. However, this definition only identifies a partial condition and, therefore, is not enough. The concept of the technological uncanny and the analysis of the Google Glass case teach us something else. Anti-mediation is a specific human experience of technology. It is the perception—through and within a technological system—of the fragility of the human identity—that is, the fragility of the border between human and nonhuman. This perception is characterized by a specific form of regression; in the anti-mediation, the human being experiences a return to matter, to an inorganic state. However, there is something more in this perception: the human being feels threatened, in the sense that they experience an invasion of the inorganic within the organic, and thus, their human identity is jeopardized. For this reason, anti-mediation is strongly connected to the concept of immunization, as I will show later.

The anti-mediation perception has deep unconscious roots. As Sloterdijk (2011–2016, Chapter 6) explains, the perception of the fragility of the boundary between human and nonhuman is connected to the theme of the "double," the doppelgänger, or the alter ego. For Sloterdijk (2011–2016), human identity is essentially divided; the human being is not a unitary whole:

> In the beginning, the accompanied animals, the humans, are surrounded by something that can never appear as a thing. They are initially the invisibly augmented, the corresponding, the encompassed and, if there is disarray, those who have been abandoned by their companions. That is why investigating humans philosophically means, first and foremost: examining paired structures, both obvious and less visible ones, those that are lived with congenial partners and those that create alliances with problematic and unattainable others. . . . Even what newer philosophers have termed "human existing" is thus no longer to be understood as the solitary individual standing out into the indeterminate openness, nor as the mortal's private suspendedness in nothingness; existing is a paired floating with the second element, whose closeness maintains the tension of the microsphere.
>
> (Sloterdijk 2011–2016, 477–478)

From birth, the human is marked by a cut, by a division. The umbilical cord is cut from the newborn baby. The link with the original environment—the placenta—is destroyed. Baby and placenta form the first microsphere, the original dual space that constitutes human existence. The placenta is the "primary companion" (Sloterdijk 2011–2016, 343), the completion of the human being. The cut of the navel and the loss of the placenta leave a physical and symbolic empty space.[3] This is the first real event in the human life, as well as the first repression. Because of this loss, the human being always feels the need to rebuild that original space, that intimate sphere. The history of culture shows this need through a series of concepts, such as the Socratic demon, the genius of the Romans, the twins, the guardian angel, and the soul. They are all imaginary figurations of the original lost "double," the placenta, and expressions of the desire to restore the intrauterine bubble. The uterine dualism of fetus/placenta is, at the same time, a non-optical[4] empathic relationship; the placenta is a container and a membrane that mediates between inside and outside and makes the existence of the fetus possible. The subject in Sloterdijk (2011–2016) is always a dual, ecological subject, in the sense that it is always dependent on its own bubble, that is, on its own ecosystem or context—we could say, following our terminology, on its *umwelt*; subject and ecosystem constitute each other.

However, because of this intimacy, the relationship with the double is ambiguous, paradoxical. On the one hand, the modern subject removes the double uterine, considering it a threat to its own identity. Cartesian doubt is a great expulsion of anything else from the original sphere. This leaves the modern subject alone. On the other hand, the modern subject is attracted by the double; it needs a double, a doppelgänger, to ensure its own identity. Following Sloterdijk (2011–2016, 435–440), the Kantian concept of "I think" is a figure of the double that observes and ensures, with its presence, the unity of the subject. Therefore, the double simultaneously allows and destroys human identity. This is also the paradox at the root of the uncanny valley phenomenon; the robot is too like the human—it is a double and, therefore, jeopardizes human identity. Technology is at the same time the memory of the original uterine duality of fetus/placenta and the symbol of the umbilical castration. Umbilical castration is terrifying and must be removed because it coincides with the rupture of the first uterine microsphere.

Therefore, the concept of anti-mediation does not simply denote the technology of which we are not aware because we can always turn our attention to that technology—we have opted not to think about it, but we can always become aware of it. Freud teaches us a different direction of thinking; the unconscious is not what we are unaware of but what we cannot, or do not want to, be aware of—that is, resistance or repression. Anti-mediation is that perception of technology that we cannot think about; we perceive it, but we cannot translate that perception into words. The assumption of this thesis is that technology plays an active role in its relationship with humans and that it can produce signs and meanings.

How do we identify an anti-mediation? Here, Freud again gives us an important methodological indication. To identify an anti-mediation, it is first necessary to analyze the human experience of a specific technology (e.g., Google Glass). Is there a disturbance in this experience? If so, it is necessary to analyze the connection between that disturbance and the mediations acting in the technology. Once the mediations have been identified, we can use the MES to verify whether that perception of a disturbance is just an effect of a suggestion. According to the MES, we (1) presuppose that that perception is an effect of a suggestion and then (2) introduce a new suggestion into the user's mind. The second suggestion is the opposite of the first.

How can we introduce a new suggestion into the user's mind? We can redesign the artifact by introducing new mediations. Once the artifact has been redesigned, we face two possibilities: the disturbance disappears, or it does not. If the disturbance is eliminated by the introduction of the second suggestion, then it was not a real perception but merely the effect of a human suggestion—a human projection on the artifact. This means that that disturbance is not an anti-mediation. If, however, the disturbance does not disappear, then that disturbance is real, and it is not a suggestive effect. Through the MES, we can distinguish suggestive perceptions of technology from non-suggestive ones because we have a key indication: a suggestion is eliminated by another suggestion of the opposite sense. In this way, we have isolated a set of non-anthropomorphic perceptions of technology.

Did we answer the original question? Only partially. The MES allows us to eliminate suggestive effects and isolate the non-suggestive and non-anthropomorphic perception(s) of technology. The MES does not allow us to identify the anti-mediation among the identified non-suggestive perceptions. However, our field of investigation has now been narrowed down, and we can operate with greater confidence. The next goal will therefore be to connect non-suggestive perceptions, including anti-mediations, to a new group of mediations and to examine the effects of this transformation.

Let us take an example. *In The Age of Surveillance Capitalism: The Fight for a Human Future at the New Frontier of Power,* Shoshana Zuboff (2019) offers a comprehensive account of the new form of economic oppression that has crept into our lives, challenging the boundless hype that has often surrounded the activities of modern technology companies. The central thesis is that, unlike industrial capitalism, which profits from exploiting natural resources and labor, surveillance capitalism profits from the capture, rendering, and analysis of behavioral data through "instrumentarian" methods that are designed to cultivate "radical indifference . . . [which is] a form of observation without witness" (379).

The surveillance capitalists found an untapped reservoir of information that their services were collecting for internal analytics and programming purposes, and they saw an opportunity: they could sell that "data exhaust" to advertisers. For them, the humans attached to that data are just accessories. Zuboff often compares the instrumentarianism of surveillance capitalism to the totalitarianism that Hannah Arendt (2017) describes in *The Origins of Totalitarianism.* Zuboff draws links between the charting of cyberspace by surveillance capitalists and Arendt's analysis of British imperialism as a forerunner of totalitarianism. Ultimately, she denies their equivalence, as totalitarianism arises in a state, whereas instrumentarianism arises in companies. As a result, Zuboff "is more concerned with surveillance by corporations than government surveillance.

Her critique of surveillance companies is often to the extent that they begin to resemble a tyrannical, authoritarian state" (Di Bella 2019).

Zuboff's analysis represents, to some extent, the generalized paranoia that is increasingly widespread in our societies. This paranoia concerns the control and use of private data by new, cutting-edge technology companies. A case in point is Amazon's voice assistant, Alexa. Numerous articles have raised the alarm about the data theft that Alexa commits to the detriment of the user. This can take the form of Alexa recording and transmitting much more data than users think it does,[5] or keeping data even when users have deleted it (or think that they have).[6] Some also claim that errors in Alexa could allow hackers to easily break into the system and steal data without getting caught.[7] In short, personal data are always in danger.

Now, is this paranoia a case of anti-mediation? What types of mediation is this perception connected to? Alexa involves two main types of mediation: *t*-background T (H/W) and alterity H → T (W) mediation. On the one hand, Alexa is an agent that defines the context in which the human agent operates; despite appearing non-invasive, Alexa connects the human agent to a planetary network of information and data exchanges, a much broader context. On the other hand, Alexa is an agent with which the human interacts to obtain certain information—the human being makes a choice and determines the behavior, or a part of the behavior, of the machine.

Now let us imagine testing the paranoia about data by transforming the existing mediations. This means that we must redesign the artifact. Therefore, we imagine redesigning Alexa on the basis of two types of mediation: hermeneutics H → (T-W) and immersion H ←→ (T/W). The relationship between human and machine is now human-centered, and this means that the human being can access a greater amount of data about the machine and its actions. Does this new design eliminate paranoid perception? Does increasing the level of symbiosis between humans and machines erase the fear of data theft? If the answer is yes, then that paranoid perception was just the effect of a suggestion. If, in contrast, the answer is no, then that perception is an anti-mediation and not the projection of a human fantasy.

Technoanalysis and Social Robotics

HRI is one of the main fields of application of technoanalysis. Johanna Seibt's work represents an interesting and important starting point for the discussion of the concept of HRI. The concept of asymmetrical relationships, developed by Seibt (2016, 2017), concerns the attempt to classify social modalities that are not symmetrical—that is, those in which the participants do not have the same attributes and abilities as one another. This operation is motivated by the thesis that we should conceive of sociality as a matter of degree; as such, according to this theory, we should abandon the idea of a dualist distinction between social and non-social interactions.

Seibt therefore focuses on describing the degrees to which a robot is capable of effectively simulating a certain type of human action (e.g., healthy action), building an expansion matrix based on this evaluation. The expansion matrix enables—for each relevant action performed by the robot—the evaluation of its effective mimetic ability in relation to the human being. Different degrees of simulation produce different reactions in the human being. Following this approach, Seibt created the concept of "sociomorphing." Seibt et al. (2020) discuss the fact that, in some experiments conducted by the Research Unit for Robophilosophy and Integrative Social Robotics of Aarhus University, what defines HRI is not only anthropomorphization but also the perception of a nonhuman social capacity, that is, sociomorphing.

In this section, I intend to (1) analyze the concept of sociomorphing by providing a critique of Seibt's point of view and (2) show that technoanalysis can offer important conceptual tools to understand the distinction between anthropomorphizing and sociomorphing through the study of suggestion and its elimination by the MES. One of the central theses of this section is that, in the study of HRI, all the interactional elements (including verbal, non-verbal, communicative, and metacommunicative) must be considered of equal importance—that is, all produce signs and meaning.

Anthropomorphizing and Sociomorphing

Seibt et al. (2020) claim that "social robotics and HRI are in need of a unified and differentiated theoretical framework where, relative to interaction context, robotic properties can be related to types of human experiences and interactive dispositions" (51). The fundamental problem in social robotics today is to develop terminological and conceptual tools that can compose a single frame of reference for the description of HRI, connecting the following:

1 Robotic design
2 Interaction contexts
3 Human experiences and interactive tendencies

It is the same question as that posed by Coeckelbergh (2011): how can we know and evaluate the relationship between human and robot? The axiological plan presupposes the gnoseological plan. We can evaluate the moral impact of a robot only if we first possess conceptual and terminological tools suitable for describing our relationship with the robot. The HRI is a specific theoretical object, which stands on the border between representing an intersubjective relationship and a relationship between a human and a thing; the robot is not another subject, but neither is it an object, an inanimate stone. The robot interacts with the human being, can communicate and even arouse unconscious emotions and reactions (Rosenthal-von der Pütten et al. 2014; Masaaki Kurosu 2020), yet it is not considered a human. The robot is a social actor, not a simple mediator; it intervenes directly in social relationships and modifies them (for the distinction between a mediator and an intermediary, see Latour 2007).

The problem of anthropomorphism is decisive and motivates the primacy of the gnoseological aspect over the axiological one. It is an obvious fact: we tend to anthropomorphize the robot, despite an awareness of its artificial nature. According to Krämer et al. (2011), the more anthropomorphic the robot is, the more human beings will tend to assume anthropomorphic interactional modalities. Hence, the following questions arise: How can we truly know HRI in all its dimensions if we constantly project an anthropomorphic image onto the robot? Do we not risk failing to give value to nonhuman aspects that could be of great importance? Do we not risk forgetting or erasing some fundamental differences between humans and robots? Are not we likely to miss some of the potential dangers inherent in that relationship? These are questions of crucial importance, as not understanding HRI could have serious social consequences. Alongside these aspects, there is the problem of the designers' biases. As Bisconti (2021a) writes, "the design of the robot, its body, is written and shaped by a socio-political competition between symbols, narratives, and mechanics of representative power. The robot-notebook is like a blank sheet where symbolic structures and conflicting narratives can collide as in a proxy-war"; therefore, "the problem that logically precedes any other must be a correct descriptive theory for a social world hybridized by interacting machines" (14–15; my translation).

This is also the premise of the argument by Seibt et al. (2020). Indeed, social robotics "still lacks standardized terminological tools for describing—comprehensively and with conceptual precision (i.e., nonmetaphorically)—what is 'going on' when humans engage in apparently 'social' interactions with 'social' robots" (54). In other words, we do not know what HRI is; we do not know how to describe it, nor do we know its real impacts, developments, and dangers. For this reason, "social robotics presents philosophy with tasks that go far beyond ethical considerations and require the involvement of theoretical disciplines of philosophy such as ontology, phenomenology, and philosophy of science" (Seibt et al. 2020, 54). The problem of description emerges above all when robots are used in particularly sensitive contexts, such as the care of the elderly.

Seibt et al. (2020) distinguish four possible approaches to the problem of description (i.e., reductionist, constructivist, fictionalist, and diversification approaches) and define its approach as a type of diversification. We can identify three general characteristics of this approach:

1 The distinction between sociability and human intelligence: Being social does not necessarily require being human.
2 The realism of capabilities: In the interaction with the robot, real social skills are attributed to the robot—not only imaginative projections.
3 The distinction between the level of expression and the level of perception in interaction: There are aspects of the interaction that are only thought and others instead that are only perceived.

Starting from these premises, Seibt et al. (2020) question the common idea that HRI always depends on the anthropomorphization of the robot. Seibt et al. (2020) instead affirm that HRI is not only the result of process of anthropomorphization or "the projection of imaginary or fictional human social capacities" (51); rather, HRI is also, or primarily, the result of the perception of nonhuman social skills, what is called *sociomorphing*. According to Seibt (2020), "there are good empirical and conceptual reasons to claim that human social interactions with robots manifest the human tendency to sociomorph robots—i.e., to attribute to them the capacities of social agents though not of human social agents" (52). Sociomorphing "can take many forms each of which is manifested in, or otherwise associated with, a type of experienced sociality" (52).

Sociomorphing can be described as (a) a form of the direct perception (b) of real nonhuman aspects in nonhuman entities and associations that is (c) capable of producing meaning in social interaction. In order to produce meaning in social interaction, the perceived aspects must be similar to human qualities and behaviors; this similarity can be evaluated according to a scale or matrix. Levels of similarity correspond to different types of simulation. Consequently, "since sociomorphing is the direct perception of actual characteristics and capacities that may resemble the characteristics and capacities of human social agency to a greater or lesser degree, sociomorphing can take many forms" (59). Furthermore, (d) sociomorphing is a preconscious perception; "it typically occurs preconsciously but may also occur consciously" (Seibt et al. 2020, 8).

Scholars affirming that social interaction with robots is based on anthropomorphization follow the traditional view of sociality, according to which social interactions presuppose consciousness and intentionality and, therefore, the ability to infer the mental states of others. However, this view can be criticized both theoretically and empirically (Seibt et al. 2020, 6–7). While anthropomorphizing is a reflective and inferential process that proceeds in only one direction (i.e., human → machine) to explain and predict the behavior of the robot through the projection of human interactional patterns, sociomorphing is the direct perception of a nonhuman social behavior; Seibt et al. (2020, 7) draw on human interactions with animals as an example of sociomorphing. Sociomorphing is therefore the perception of a form of sociality that does not fit, or does not fully fit, into human interaction schemes.

Now, although Seibt et al. (2020) have the merit of introducing the distinction between anthropomorphizing and sociomorphing, the theoretical framework they use to describe and explain this distinction can be contested. Seibt et al. (2020) refer to the Ontology of Asymmetric Social Interactions (OASIS), a type of approach that is also developed in Seibt (2017). My main criticism of the OASIS is that it reduces every aspect of HRI to simulation by implicitly adopting an anthropocentric perspective. In the OASIS, the lead of the HRI is the human—it is the human who admits the participation of the robot in the interaction.

The main problem the OASIS seeks to solve is "how a certain robot participates in an interaction with a human" (Seibt et al. 2020, 136). The relationship is always asymmetrical, so it is "an interaction where one interaction partner simulates (her, his, its by her) agentive contributions" (136). There is the human being, the only source of meaning, and there is the robot, which imitates the human—a dualistic situation. The problem lies in the concept of simulation, which contradicts that of sociomorphing. My objection to Seibt is that solving the description problem through the OASIS and the concept of simulation could erase aspects of HRI that cannot be reduced to simulation—that is, those aspects that are shaped by the machine itself, such as situations in which the machine has an effect on the human being, or the human being changes their behavior due to the presence or action of the machine. This criticism is also shared by Bisconti (2021a).

According to Seibt (2018), "a simulation is nothing else but the occurrence of a similar but simpler process with sufficiently similar effects" (137). In logical terms, simulation is understood as a kind of deviation from a basic model, what Seibt (2018) calls a "realization" of an action. The realization is defined as follows: "For any process type *A*, process system S realizes *A* iff S generates a process *p* that is an instance of *A*" (137). Seibt (2018) therefore distinguishes five types of deviations: (1) functional replication, (2) imitation, (3) mimicking, (4) displaying, and (5) approximating: "Taking both human beings and robots as process systems, we can describe what the robot 'does' as the replication, imitation, mimicking, display, and approximation of the doing of a human process system" (Seibt 2018, 138–139). Furthermore, to "talk about robots 'acting' is justified *only if we thereby understand that we implicitly refer to (some degree of) a simulation or 'manner of performance'*" (Seibt 2018, 139; emphasis added). Only the aspects of HRI that fall within the levels of similarity displayed by Seibt et al. (2020) in the matrix (on pages 139–140) can be admitted into the description of the HRI. Any other aspect is not considered relevant.

This position, in my opinion, misses the point. The theoretical model of ANT gives us much more compelling tools to describe HRI without having to adopt the point of view of simulation, which is still too anthropocentric. I argue that in the description of HRI, it is necessary to consider all the interactional aspects as forms of agency—not only those that adapt to the human model. In an interactional context, anything is interaction and anything can influence interaction. I therefore propose to substitute the concept of simulation with that of negotiation.

My second objection to Seibt is that we cannot take one–one relationships as a model to evaluate situations involving multiple humans and multiple robots, that is, complex interactions. Seibt assumes that what happens in complex interactions can be deduced from simple one–one interactions in laboratory settings. I claim that while the simulation model works in the laboratory, it does not work when we have to describe complex and unpredictable interactions.

Let me be clearer on this using an example: a robot comes toward me and greets me. Why is that greeting a simulation? What is it a simulation of? Who is giving the greeting, the robot or the software that makes it work? There is no conscious intention in the software, so that greeting is not an intentional act; the robot recognizes a situation (i.e., someone entering the room) and then applies a procedure (i.e., giving the greeting). Where is the simulation? The simulation lies in the eyes of the human agent who receives the greeting; it is the human agent who expects to receive a greeting upon entering a room, who can explain the presence of that robot in that room as something that must welcome passersby, and who then interprets it in that way (i.e., by anthropomorphizing the robot). Furthermore, the simulation lies in the team of technicians who designed and created the robot to simulate a certain behavior, such as greeting people who enter the room.

What if the intention of the technicians had been different? What if raising an arm and saying the word "hello" were the result of a technical error, a bug in the robot's software? Let us say that the software that runs in the robot was originally meant to allow the robot to illustrate an architectural project; a malfunction produced a bug, making the robot now only able to raise its arm and say "hello." The bug produces a crisis. Is the robot still simulating? What is it simulating—the action prepared by the designers or the one envisaged by the human agent it meets? When asked, "Does the robot simulate?," the designers would respond in the opposite way to the human agent who meets the robot.

Let us take another case. Due to a bug in the software, the robot is not only unable to illustrate an architectural design (which it was supposed to do), but it cannot even say the word "hello"—meaning that its action is limited to raising its arm, in a rather fast and violent way. Let us say that a human agent, entering the room, interprets the raising of the robot's arm as a provocation—for example, as a Nazi salute. Another human agent who arrives at that moment interprets the raising of the robot's arm as a request for help; the human agent thinks that the robot is in danger, threatened by a person who wants to destroy it. I ask the same question as before: Is the robot simulating? What does it simulate? Is simulation really the central issue here?

What I mean to illustrate with these examples is that, in a complex and unscheduled interaction situation, *everything can be a simulation of everything*. Building the Seibt matrix would be very complex, if not reductive. Instead, I propose to go beyond the simulation model. The robot is like a text to be interpreted; each human agent can assign a different interpretation—even opposing interpretations. Furthermore, like a text, the robot can, in turn, produce new meanings and influence human beings, or even induce them to perform other actions. As Eco (2020) explains, the text creates its reader—or better, the text and reader cooperate and construct each other. For these reasons, I believe that ANT is a better model for describing and explaining HRI and sociomorphing. Anthropomorphizing and sociomorphing are not preconditions but the results of a previous negotiation process.

My third objection to Seibst's thesis is that the adoption of the OASIS as the theoretical frame of reference makes it impossible to conceive the notion of sociomorphing in all its breadth. There are different levels of sociomorphing—some conscious, some preconscious, and some unconscious. Seibt et al. (2020) do not consider the unconscious level. I claim that sociomorphing first includes those reactions produced by the robot in the human that is unconscious. The more the robot is able to reproduce human—human interaction (HHI)—in other words, the more anthropomorphic it is—the more the interactional aspects that do not correspond to HHI are perceived and experienced at an unconscious level. Several studies support this thesis.

For instance, Bainbridge et al. (2008) explore how a robot's physical or virtual presence affects the unconscious human perception of the robot as a social partner: "The level of a robot's presence affects some variables in human-robot interaction that should be important to consider when creating a human-robotic social interaction" (6). The paper describes an experiment in which subjects collaborated on simple tasks involving moving books with either a physically present humanoid robot or a video-displayed robot. Each task examined a single aspect of interaction: greetings, cooperation, trust, and personal space. The subjects readily cooperated with the robot in both conditions. However, the subjects were more likely to fulfill an unusual instruction and to afford greater personal space to the robot in the physical condition than in the video-displayed condition.

> Although subjects enjoyed interacting with both the physical robot and the video-displayed robot, they clearly gave the physically present robot more personal space. Personal space could be interpreted

> as a variable of respect, as humans give personal space to those they are unfamiliar with but respect as human.
>
> (6)

A study by Shimada et al. (2006) uses an android that has a human-like appearance to analyze the impact of the presence of the robot on the unconscious behavior of the human gaze. The main thesis is that "unconscious reactions are useful in investigating the human likeness of the android" (162). The authors presuppose that humans show unconscious behaviors when interacting with another human, and they expect the same behaviors to be present when humans interact with a very human-like robot. "We can change appearance and motion to study how the unconscious behavior of the human changes. In this way, we explore the uncanny valley" (157).

Usually, eye movements are used to send social signals during a conversation—this is a form of metacommunication. When thinking about the answer to a question, humans tend to look away from the questioner. The change of the interlocutor determines a change of gaze. The authors found that

> the subject changes gaze to the left of the face a longer time in case of a human or android questioner. The subject changes gaze to look down from the face in the case of a mechanical robot questioner. There is a significant difference between these two behaviors.
>
> (Shimada et al. 2006, 157)

Therefore, the android questioner is treated in the same way as the human questioner, while the mechanical robot questioner is treated differently. In both cases, the questioner is aware of the difference of the machine (i.e., that the interlocutor is a machine). When the robot is not anthropomorphic, there is no way to "bypass" the nonhuman aspects; however, when the robot is anthropomorphic, the nonhuman aspects are "bypassed" on an unconscious level. In other words, when the robot is anthropomorphic, the non-HHI aspects (i.e., sociomorphing) are perceived on the unconscious level. When the robot is not anthropomorphic, both the HHI and non-HHI aspects are conscious at a minimum level.

Horstmann et al. (2018) instead consider the interesting problem of switching off robots. In an experimental laboratory study, people were given the choice of whether to switch off a robot with which they had just interacted. The style of the interaction was either social (i.e., mimicking human behavior) or functional (i.e., displaying machine-like behavior). Moreover, the robot either voiced an objection to being switched off or remained silent. The results show that participants let the robot stay switched on when the robot voiced an objection to being switched off. After a functional interaction, people evaluated the robot to be less likeable than after a social interaction, which in turn led to a reduced stress experience after switching it off.

Despite this, individuals hesitated the longest to switch a robot off when they had experienced a functional interaction in combination with an objecting robot. This unexpected result might be due to the fact that the impression people had formed based on the task-focused behavior of the robot conflicted with the emotional nature of the objection. Therefore,

> people who liked the robot after the social interaction better experienced more stress, probably because they were more affected by the switching off situation. Most likely, they developed something like an affectionate bond with the robot and thus switching it off was challenging and influenced their emotional state.
>
> (Horstmann et al. 2018, 18)

The paper therefore illustrates a very interesting situation: sociomorphic perception is an emotion caused by a nonhuman—in this case, an anthropomorphic robot. An unconscious sociomorphic perception that influences the agency of the human corresponds to the conscious anthropomorphic projection. A non-verbal, non-HHI element is able to produce meaning in that situation.

A very similar approach is that of Giannopulu and Watanabe (2015). They analyze emotion, language, and (un)consciousness in children aged 6 and 9 years via listener–speaker communication. The speaker was always a child, and the listener was either a human interactor or a robot interactor (i.e., a small robot that reacted to speech expressions only by nodding). Unconscious non-verbal emotional expression associated with physiological data (i.e., heart rate) as well as conscious processes related to behavioral data (i.e., the number of nouns and verbs used in addition to reported feelings) were considered. The paper shows that (1) the heart rate was higher in children aged 6 years old than for children aged 9 years old when the interactor was the robot and (2) the number of words (i.e., nouns and verbs) expressed by both age groups was higher when the interactor was a human. However, the number of words was lower for the children aged 6 years than it was for the children aged 9 years. "Unconscious nonverbal emotional expression differs between the children aged 6 and 9 years when the interactor is the robot. Children aged 6 years manifested more nonverbal emotional expression in the presence of the interactor robot than the children aged 9 years" (Giannopulu and Watanabe 2015, 268). Even though a difference in consciousness exists between the two groups, "everything happens as if the interactor Robot would allow children to elaborate a multivariate equation encoding and conceptualizing within their brain, and externalizing into unconscious nonverbal emotional behavior i.e., automatic activity" (260).

This paper demonstrates how negotiation takes place between humans and robots. In the beginning, there is no clear distinction between sociomorphing and anthropomorphizing. These two aspects derive from the interaction between the actants, that is, the children and interactors. It is from this interaction that aspects of anthropomorphizing (e.g., conscious speaking) and aspects of sociomorphing (e.g., unconscious reactions) are defined. A process of negotiation and mutual interpretation then takes place—not the simulation of a preconceived model.

Suggestion and HRI

I now want to develop a different descriptive method of HRI—one that is not based on simulation but on negotiation and ANT. We must first recognize that even in the study of HRI, there is a problem related to the observer: to what extent does the observer disturb the observed HRI? In Chapter 1, I explained that in psychoanalysis, the MES is used to distinguish between changes that are due to suggestion and changes that are not due to suggestion. This is what distinguishes the Freudian method: the use of suggestion as a method of control. I now propose to use the same method to study HRI. Therefore, when we face a context involving HRI, to understand which aspects derive from the anthropomorphizing tendency and which do not, we must first induce a change in the context. Then, we apply the MES, meaning that we introduce a suggestion in the opposite sense; given the very nature of suggestion, one suggestion is always eliminated by another suggestion in the opposite direction (see Chapter 1). If the change vanishes, then that change was due to suggestion, that is, to the observer's projection on the HRI. If that change persists, then it does not depend only on the observer's projection on the HRI. In other words, we study HRI as if it were a black box in which we introduce inputs and evaluate outputs through the MES.

What does it mean, in this case, to "introduce a change"? If the psychoanalyst speaks in order to communicate their construction to the patient, what does the technoanalyst do? In answering this question, the theory of mediation outlined in the previous sections can help us. The psychoanalyst collects and studies the material provided by the patient and formulates an etiological hypothesis; this is the theoretical level of investigation. The psychoanalyst then communicates their hypothesis to the patient in the manner deemed most appropriate; this is the practical intervention.

The technoanalyst works in a very similar way: first, they collect and study the material provided by the HRI, and then they formulate their hypothesis of description, identifying the mediations and anti-mediations involved. Therefore, the hypothesis concerns the mediations and anti-mediations involved, their connections, and the nature of the information that circulates in them. The technoanalyst then introduces a change in the HRI[8]; this change is connected to the hypothesis because it intends to confirm or deny the hypothesis. Therefore, the change introduced is a new form of mediation or a set of connected mediations. Once the change is introduced, the technoanalyst observes its effects on the HRI. As explained in the previous sections, adopting the taxonomy of mediations defined by post-phenomenology does not mean abandoning ANT. If the change introduced in the HRI is the result of the suggestion and projection of the technoanalyst, it will be automatically eliminated by a t-mediation, meaning a mediation in the opposite direction. Consequently, the mediation—or the set of mediations—that resists the test of t-mediation is not produced by human suggestion.

We have therefore achieved a first negative result. In each situation, we know how to eliminate any projection of the technoanalyst observing the HRI and, therefore, how to isolate a core of mediations and t-mediations that cannot be assimilated into the observer's anthropomorphizing tendency. Now, we must confront the following question: how do we isolate the non-anthropomorphic element of interaction? Once again, the Freudian method can help. Suggestion is not just about the relationship between the observer and the HRI. There is a second form of suggestion—one that intervenes in the relationship between the human and robot inside the "black box." I argue that anthropomorphizing is a form of suggestion that humans project onto the machine; that is, it is a set of stimuli that, consciously or unconsciously, aims to induce a certain reaction or behavior in the receiving entity (i.e., the machine, in this case).

The technoanalyst must therefore (1) identify the essential mediations between humans and the machine outside the "black box" (i.e., those mediations that do not belong to the HRI but to the projection of the technoanalyst themself, as mentioned earlier; see Pasquale 2015) and (2) identify which of the isolated mediations in the first phase are the result of another human anthropomorphizing (i.e., the projection of the humans onto the machine inside the "black box," or inside the HRI itself). Also, in this second case, the technoanalyst must analyze the material and formulate their hypotheses before introducing new mediations in the HRI. T-mediations are systematically opposed to the new mediations, following the MES. The changes affect the setting of the HRI and/or the design of the robot. The mediations that resist even in this second phase are sociomorphic mediations, that is, non-anthropomorphic forms of sociality. In this non-anthropomorphic relational space is the true HRI. The robot is not only a mass of human projections but also an authentic social agent that is capable of certain types of mediation that structure the network in which it is inserted.

My central thesis is that every aspect of HRI must be considered significant—not just aspects that are similar to HHI, as much of the existing literature tends to suggest (Duffy 2003; Kiesler et al. 2008; Luria et al. 2019; Yuan and Dennis 2019). *Every interactional aspect is important.*

As Latour also points out, we must consider every agency and every intentionality, as well as every capacity for signification; "existence and signification are synonyms. As long as they are acting, agents signify. This is why their signification can be followed, pursued, captured, translated, formulated in language" (Latour 2017, 70). The non-HHI aspects in the behavior of the robot are not to be discarded and eliminated. Instead, they are different forms of agency, capable of producing different meanings and of reconfiguring human sociality. The non-HHI aspects indicate "new interactional settings, qualitatively different from those existing in human relationships" (Bisconti 2021b, 4).

The methodology just described must be tested and developed through future experiments. However, the literature already confirms its consistency and usefulness. Serholt (2018), for example, takes a critical look at children's interactions with a robotic tutor in an educational setting. The robotic tutor was developed as part of an interdisciplinary EU project called EMOTE (EMbOdied-perceptive Tutors for Empathy-based learning) and was implemented in a school in Sweden. The robot was designed "to tutor students in activities based on the syllabi for geography and social studies for elementary education, seeking to offer educational value to participating schools, in addition to the scientific value of studying CRI [children—robot interaction]" (Serholt 2018, 250). The paper analyzes the problems that can arise in the interaction between students and robots in the classroom and, therefore, how breakdowns occur in this relationship. Teaching is a very interesting situation because it allows us to observe different forms of breakdowns and, accordingly, different resolution strategies implemented by humans or robots. I do not want to analyze these aspects, on which there is literature already available (Sharp et al. 2007; Iacovides et al. 2015); instead, I want to focus on the results of Serholt's (2018) study.

The technical setup consisted of a robotic tutor (NAO T14 from SoftBank) and an interactive touchtable for the educational activities (55" touch-sensitive interactive display from MultiTaction). Also, a Microsoft Kinect 2.0 and a web camera connected to OKAO Vision software were used for analyzing the emotional states of the students. Two video cameras (a camcorder and a GoPro) as well as a Zoom H4N with a connected external overhead microphone were used to record the sessions. The robot could realize two types of interactional elements: verbal elements (i.e., text and voice) and non-verbal elements (i.e., gestures and displaying different eye colors). In terms of perception, the robot was equipped with face tracking, facial expression recognition to determine emotional states, and the functionality to register and interpret students' input on the interactive touchtable. Two activities were carried out with the robot, one of which was an individual map-reading scenario and the other of which was a sustainability scenario in which students participated in pairs (Serholt 2018, 253).

This study is important because it shows that breakdowns in CRI—what we previously called *anti-mediation*—occur when the expectations and projections of the children with respect to the robot are disregarded—that is, when the robot does not behave as it should. Children are perfectly aware that they have a robot in front of them, yet they expect human-like behavior.

> Fooling children into believing that a robot understands them does not offer much consolation when there is a structured task to be carried out, or a fruitful collaboration to be upheld. CRI breaks down when expectations go unmet, i.e., that robots should have humanlike perception and communication abilities.
>
> (Serholt 2018, 263)

The breakdown therefore arises from the perception of nonhuman interactional elements that cannot be repressed.

Serholt (2018) claims that "trouble in interaction is not unique for CRI or HRI. It also occurs in HHI all the time. The difference is that HHI (or teacher-student interaction, if you will) is not scripted; instead, interaction between human subjects is eloquently negotiated" (263). However, a form of negotiation also takes place in CRI and HRI. Serholt (2018), in fact, shows that the perception of nonhuman interactional elements does not necessarily lead only to a breakdown in CRI but also to strategies for redefining the relationship itself that can have positive effects on learning. In this case, the elimination of human suggestive effects would have led to the description of sociomorphing and to an understanding of the dialectical nature of the relationships between anthropomorphizing and sociomorphing—in other words, to the importance of non-HHI interactional elements.

The themes of the breakdown and reconfiguration of HRI are also at the core of Bisconti (2021b), who analyzes the fundamental question of metacommunication in HRI. Drawing upon Watzlawick's argument (Watzlawick et al. 2011), Bisconti (2021b) states that the human being interacts with the robot both on a communicative and metacommunicative level, while the robot is capable of interacting only on a communicative level. Consequently, on the metacommunicative level, the human being is forced to interpret; "humans cannot *not* interpret communication on the metacommunicative level," and therefore, "*any type of response the robot provides at the message level will also be interpreted on the metacommunicative one*" (Bisconti 2021b, 9; emphasis added). For example, humans attribute intentionality to non-verbal cues, such as eye movement, personal space, the approaching direction, the velocity of movements, and facial mimicry. "Every nonverbal cue conveys a precise metacommunicative level that acquires meaning only in relation to the other elements of the interactional system" (9). In other words, the non-verbal aspects of the interaction play a metacommunicative role; the human being tends to interpret the non-verbal elements of the robot as metacommunicative elements, despite the absence of the robot's understanding of or desire for real significance on this level.

The level of metacommunication is fundamental because it is the one on which the negotiation between the actants in the interaction takes place. Now, in HRI, the human being relies "on the presumption that implicit metacommunicative content is understood correctly by the recipient, who is supposed to respond consistently" (Bisconti 2021b, 11). Furthermore, "because of the systemic nature of interactions, every interactional element retroactively shapes the meaning of past interactions" (11). Interacting with a robot that does not understand the metacommunicative level forces us to redefine the metacommunicative level and its systematic dimension. The robot can confirm or deny human expectations, according to different degrees of intensity, and this has systemic consequences. Bisconti (2021b) therefore shows that the inability to understand the metacommunication level does not make the robot unable to act or make sense in an HRI. If we assume the Latourian equation that agency = the ability to produce meaning, then we must admit that the robot acts and therefore produces meaning and communicates due to the interaction itself, which are the characteristics of a communicative situation.

I think that both papers we have analyzed confirm the action of suggestion between humans and robots. Furthermore, studies on the psychological effects of robots on humans (especially Bisconti 2021b; Massa et al. 2022) show that suggestion can also go from robots to humans. The robot can produce a certain form of psychological pressure on the human, prompting the human to behave in one way and not another. The technoanalyst must therefore also act on this type of suggestion, using the MES reinterpreted through the post-phenomenological theory of mediation. This is a third form of mediation, which is added to the two previously mentioned: (1) that from the observer to the HRI *outside* the HRI and (2) that from the human to the robot *within* the HRI.

Therefore, following ANT, we must affirm that anthropomorphizing and sociomorphing are the final outcome of a much more complex translation process. In saying this, I am not magically attributing intentionality to the robot. Instead, I am distributing intentionality among the different actants in the network. I claim that there is no sociomorphing and anthropomorphizing; there is instead anthropomorphizing and robotmorphing, which are two products of a more original sociomorphing. This morphing is (a) contextual, meaning that it is always particular—it depends on the umwelts involved; (b) chaotic, devoid of a pre-established order; (c) translation, or an exchange of quality between actants; and, therefore, (d) implies a negotiation. There is no human and robot; there is always the *human–robot*, the cyborg. Human and robot are always the result of a negotiation.

Satake et al. (2009) showed a concrete example of negotiation through an experiment that involved a robot actively approaching people in a shopping mall to offer various services. Negotiation is carried out by the robot, which also defines the conditions of the interaction, not the humans. The study demonstrated the difficulty of giving the robot the ability to initiate an interaction with a human being in an unstructured situation. On the basis of the study's results, at least four moments can be distinguished: (1) the phase of the approach (metacommunicative level), (2) the beginning of the conversation (communicative level), (3) the establishment of the social relationship (metacommunicative level + communicative level), and (4) the solution of possible interactional problems. The first phase is the most complex, with the robot analyzing the behaviors of a human and establishing whether this individual can be contacted—the human does nothing. In this phase, the robot, not the human, anthropomorphizes. It determines whether an actant has the right characteristics to be defined and treated as a human being. At the same time, the human can respond positively or negatively on the grounds of his/her perceptions regarding the robot. A positive reaction implies the achievement of balance between (1) a certain degree of anthropomorphizing (with the robot accepted as an active interlocutor similar to any other human) and (2) a certain degree of robot morphing (wherein the robot is recognized as a robot, a machine, with all the communicative limitations that this entails and that can be justified; it is the level that Seibt et al. 2020 calls sociomorphing).

Interaction begins only when the robot and the human being have reached an agreement on the metacommunicative level. The challenge is that this level is never the same; actants (robots and humans) analyze substantial data differently in deciding how to manage a situation (whether to initiate an interaction). An agreement is possible if and only if a few conditions and variables are satisfied at the same time. This basic metacommunicative agreement determines reciprocal strategies. For example, a human being decides not to interact with a robot, rendering the negotiation a failure, and the human chooses not to admit the robot into his/her interactional space. Alternatively, the human decides to admit the robot into his/her interactional space but only recognizes it as a robot and regards communication as having no meaning (i.e., the reaction of rejection).

In conclusion, technoanalysis applied to social robotics intends (a) to highlight the role of suggestion in HRI; (b) to propose a methodology to eliminate suggestion from HRI and to describe the sociomorphing in a more objective way; and (c) to give all aspects of HRI (human and nonhuman, verbal and non-verbal, and communicative and metacommunicative) the same importance. I believe that the literature discussed demonstrates the plausibility of such a research line. Finally, technoanalysis involves an operational intervention; it improves HRI, making it more balanced through design solutions that are based on the theory of mediation.

Interaction Failures

What types of anti-mediation are identified in technoanalysis studies on HRI? Here, I mention the case of interaction failures (IFs), or those situations in which interactions between humans and robots fall into crisis. It remains unclear when people perceive and resolve robot failures, how robots communicate failures, how failures influence people's perceptions and feelings toward robots, and how these effects can be mitigated. "Research suggests that the relationship between symptoms and cause of failure is often not clear even to trained roboticists" (Honig and Oron-Gilad 2018, 2; see also Steinbauer 2013). Customer support "also becomes costly when users are unable to differentiate between technical errors (software bugs or hardware failures) and problems resulting from improper use or unrealistic expectations" (Honig and Oron-Gilad 2018, 3).

There are many IF taxonomies (see Laprie 1995; Ross et al. 2004; Giuliani et al. 2015; Carlson and Murphy 2005). Honing and Oron-Gilad (2018) distinguished between technical failures and IFs. The former are caused by technical problems in hardware or software. Problems in software are, in turn, classified into three broad groups: design, communication, and processing failures. Conversely, IFs are "problems that arise from uncertainties in the interaction with the environment, other agents, and humans. These include social norm violations and various types of human errors." (Honing and Oron-Gilad 2018, 3). On the basis of these definitions, there are three large groups of IFs: violations of social norms, human errors, and violations from the environment and other agents. Human errors can take the form of mistakes, slips, lapses, and deliberate violations. An IF can also be composed of several types of failures.

A survey of the literature, which, given the vastness of the topic, cannot be exhaustive, revealed a difficulty in distinguishing technical failures from IFs and, therefore, in identifying the conditions and symptoms of the latter. No study has considered the problem of the observer effect on HRI and IFs, and most studies view IFs only as a problem to be solved instead of an issue that can considerably illuminate the types of mediation and anti-mediation taking place in HRI. Technoanalysis can help isolate fundamental mediations in HRI and thus pinpoint, communicate, and solve actual IFs. This method involves interpreting IFs as the expressions of an older epiphylogenetic layer that is no longer accessible to the conscience of a human user but continues to "press" on newer epiphylogenetic layers.

Anti-Mediation and Immunization

At this point, a careful observer might ask themself the following: what is the relationship, if any, between our theory of mediation and psychoanalysis? To what extent is our investigation still psychoanalytic? What is left of Freud? What does the concept of anti-mediation have to do with that of repression? Where is the connection?

Psychoanalysis is, in my eyes, "still the most interesting interpersonal practice of closeness in the modern world" (Sloterdijk 2011–2016, 298). From this point of view, psychoanalysis is the only true exploration of the sphere of intimacy and, above all, of the sphere of originary intimacy—that is, the bipolar sphere of closeness and intimacy, which exists in the relationship between the fetus and the mother in the uterus. The final goal of psychoanalytical constructions is the knowledge of this fundamental and original experience that conditions the entire life of the human subject. Following Macho's (2011) critique of Freud, Sloterdijk questions the Freudian theory of psychic phases by accusing it of being based on an objective conceptuality that is inadequate to explain the uterine experience. To understand the sphere of uterine intimacy—the first real human space that defines each subsequent form of identification—we need a theory of mediation. The reason for this is that the experience of the fetus in the uterus is the experience of a mediation, of a relationship—but not of a relationship between a separate subject and an object placed one in front of the other.

The general theme of *Spheres I* (Sloterdijk 2011–2016) can be defined as an archeology of intimacy. Sloterdijk investigates the constitution of the individual from the stage of gestation in the maternal womb. He constitutes, using the theoretical tools of depth psychology, medicine, and the history of culture, a philosophical anthropology that does not naively escape comparison with modern sciences but that uses this comparison to elaborate a new complex theory of human subjectivity, from which to start reinterpreting the history of Western philosophy and civilizations.

Sloterdijk draws on Macho's concept of *nobjects*. He defines *nobjects* as realities that displace the observer, placing in front of the observer something that does not yet have an objective presence—in other words, objects that are not given or realities that abolish the subject/object division because they precede it. The first nobject Sloterdijk deals with—one could even go so far as to argue that, in all his philosophical speculation, he deals with nothing else—is the mother as a receptacle of intimacy, as pure interiority, as a vulva, as a cave, as a door between pre-originary interior and exteriority, as the only reality that is properly given to us. The fetus is also a nobject, which is not yet a subject but which cannot be defined as an object either. Starting from these assumptions, Sloterdijk criticizes the three phases that, according to Freudian psychoanalysis, give the description of early relationships (i.e., oral, anal, and genital). He thinks that this theory is undermined from the outset by the lack of recognition of the mediating nature of the nobjects.

Now, Sloterdijk (2011–2016) distinguishes three pre-oral phases that belong to uterine life: (1) fetal cohabitation, in which the fundamental media are blood and amniotic fluid that protects and nourishes the fetus;

(2) the psycho-acoustic initiation of the fetus, or the relationship of the fetus with the sounds coming from the mother; and (3) the respiratory phase, or the moment of detachment from the mother and the transition from immersion in amniotic fluid to immersion in air, another medium. These three stages are primitive forms of mediation. Mother, uterus, fetus, umbilical cord, amniotic fluid, and blood—they are all nobjects. In Sloterdijk's view, globalization and the technology it implies are nothing more than an extension of these three initial spheres, of these places of existence—the search for and recovery of that form of closeness and protection. The uterus is the first experimentable *différance* (Derrida 1974).

What does this mean for technoanalysis? We can trace a connection between the theory of post-phenomenological mediation and Sloterdijk's spherology. Every form of technological mediation can be considered a repetition of uterine mediations and protections, just as anti-remedies are "distortions of participation" (Sloterdijk 2011–2016, 298). Technoanalysis, therefore, follows the path of Sloterdijk's spherology in the opposite way. What else is technology if not the precise embodiment of the idea that it is possible to overcome the difference between being outside (of the uterus) and being inside (of the uterus), in a unit placed at a higher level? Is not AI the most powerful symbol of this idea—an external becoming internal, and vice versa? Is not AI the most powerful attempt to reconstruct the uterine ecosystem? Therefore, identifying technological mediations is not enough; to identify the anti-mediations, it is necessary to interpret the technological mediations that are active in the artifact we are analyzing according to Sloterdijk's three phases, which are forms of mediation too. How does embodiment mediation reproduce fetal cohabitation? What are the internal mediations between these two forms of mediation? Can we find a language and a conceptual architecture common to both?

Let us try to understand this point better by extending the analysis of the three original pre-oral phases. The first pre-oral phase is a fetal cohabitative phase in which there is the experience of the sensory presence of liquids, bodies, and the limits of the uterine cavern. Here, as a precursor of the reality that will later become the world, there is an intermediate fluidic realm, that is, the maternal prenatal uterine environment. This first phase, according to Sloterdijk, will recur continuously, overwhelmingly, as living—understood as being in space, building space, and inhabiting a space humanized (and humanizing); this will be the fundamental characteristic of the human being. For Sloterdijk, in fact, the human will be nothing other than that animal that creates and inhabits a space. Volumes II and III of *Spheres* are mainly dedicated to the explanation—on the supra-individual and intersubjective historical, philosophical, and cultural levels—of this concept. The immersion of the fetus in amniotic fluid and blood, and the relationship with the placenta, are the other fundamental characteristics of this stage.

The second pre-oral phase is the psycho-acoustic initiation of the fetus into the uterine sound world, in which attention is paid to the importance of the voice as an umbilical cord that still unites, after delivery, the newborn with the mother, and that is the germ of all future communication. Sloterdijk dedicates Chapter (VII) of *Spheres I* to the deepening of the relationship of the fetus with music and hearing in general. The last preoral stage is the respiratory phase. As seen, the first real experience of the subject *in utero* is that of immersion in a fluid medium within concrete spatial boundaries. Immersion can also occur, however, in an invisible medium that is active, such as the Internet.

From this point of view, the technological artifact is a set of objects that reproduce and extend this topology of being through which humans define themselves and inhabit the world. In addition, every mediation described by postphenomenology and their connections can be interpreted according to each of these three phases.

Through the theory of the pre-oral phases, Sloterdijk helps us understand another crucial aspect of technology: immunization, which refers to the ability to protect and preserve the human umwelt. According to Sloterdijk, the reference to the uterine world is present in every relationship with technology. What we ask of the artifact is to first ensure our safety, protect us from danger, and define a safe space from which we can resist to protect the enclosure. This is an original, unconscious request that is part of the organic bipolarity typical of uterine life, which is the first form of the umwelt. For Sloterdijk, uterine immunization comes from a relationship with the other—the mother—which emerges from bipolarity. From this point of view, AI is a symbolization of intrauterine psychic bipolarity, as are doppelgängers, the so-called doubles and soulmates in mythologies (see Sloterdijk 2011–2016, 362–69).

Technology, immunity, and intimacy are three key concepts that cannot be separated. This connection also emerged during the COVID-19 pandemic with the production and dissemination of vaccines. Immunization is a key issue of biopolitics today. As Esposito (2022) points out, community and immunity are two inseparable concepts. In other words, there is no community without tools that ensure our immunity. No social body would have resisted the conflicts that it came across over time without a protective system capable of ensuring its permanence, and the same can be said of the human body. Technology is the first tool through which immunity is sought. Even before science, it is technology that must satisfy that need, and it is in technology (in the form of the vaccine, for example) that the two sides of the immunity problem, the juridical-political one and the medical-biological one, converge. Immunization is "the secret name of civilization" (Esposito 2022, 13; my translation). Sloterdijk offers a positive interpretation of immunity as entailing protection, intimacy, maternal affection, and the search for balance. From this point of view, and also from the perspective of ANT, immunity corresponds to the strategies implemented by the collective to respond to internal and external threats that prevent it from stabilizing.

The reference to the connection between technology and immunity gives us the conceptual resources to better understand the concept of anti-mediation. Immunization is the key to understanding technology as an attempt to return to uterine duality. Anti-mediation is therefore the dissolution of the original pact with technology, the loss of the promised immunity, and therefore the rupture of the envelope and the psychosis of the absolute outside. Anti-mediation is the essence of every psychosis as a "spheric catastrophe" (Sloterdijk 2011–2016, 329) and of modernity itself. Anti-mediation produces lonely, weak, hysterical human subjects who are unaware of their "where" and their own roots. For this reason, anti-mediation is a concept that lies at the intersection of post-phenomenology, phenomenology, spherology, and political theory.

Conclusions

In this chapter, I analyzed the theory of mediation starting from post-phenomenology. I have shown that it is compatible with ANT, in the sense that, beyond its radical diversity on the ontological level, the post-phenomenological theory of mediation can be used in the context of ANT on the methodological level to recognize and describe the different kinds of association and translation in a network. I then developed a descriptive method based on the concepts of mediation and anti-mediation and showed its applicability to some case studies, including Google Glass and the Metaverse. Finally, I showed the application of this method to the field of social robotics in connection with the MES. I believe that the study of suggestion and its effects in HRI is of fundamental importance. The MES is the most convincing tool we have for studying and eliminating suggestion and its effects.

Notes

1 In writing this part of the chapter, comments on the following videos were reviewed:

- "What Happened to Google Glass?" (www.youtube.com/watch?v=2a-14kmv1zA)
- "Google Glass How-To: Getting Started" (www.youtube.com/watch?v=4EvNxWhskf8)
- "Introduction to Google Glass" (www.youtube.com/watch?v=cAediAS9ADM)
- "Project Glass: Live Demo At Google I/O" (www.youtube.com/watch?v=D7TB8b2t3QE)[2]

2 See www.breakinglatest.news/world/facebook-changes-its-name-it-is-now-called-meta-and-aims-at-virtual-reality/.

3 See considerations on the relationship between the navel and the unconscious; the navel is the symbol of an event that no one remembers (Sloterdijk 2011, 200–201).
4 It is important to note Sloterdijk's (2011, 439–440) critique of Lacan and the concept of the mirror stage.
5 See www.pcmag.com/news/amazons-alexa-collects-more-of-your-data-than-any-other-smart-assistant.
6 See www.theverge.com/2019/7/3/20681423/amazon-alexa-echo-chris-coons-data-transcripts-recording-privacy.
7 See www.theverge.com/2019/7/3/20681423/amazon-alexa-echo-chris-coons-data-transcripts-recording-privacy and www.bbc.com/news/technology-53770778.
8 This is done using both qualitative and quantitative methods, such as observations, interviews, focus groups, and the study of the data provided by the experiment.

4 Looking Through Replika

How to Psychoanalyze an AI Chatbot[1]

An article published on 30 September 2020, in the Italian newspaper "Corriere della sera" denounced a smartphone application that incited the killing of people. The application under accusation is Replika, which is based on AI; it learns to recognize patterns and predict certain behaviors and situations from the data provided and guides the user to act accordingly. Launched in March 2019, Replika is currently used by approximately seven million people.

Replika has been defined by its creators as "the AI companion who cares." This chatbot aims to help people psychologically through conversations. Using a machine learning system, Replika learns to recognize feelings, memories, dreams, and thoughts and tries to understand its users and support them. Without providing professional psychological help, Replika can become your best friend, boyfriend, girlfriend, or mentor, as well as help you get out of loneliness, a bad day, or just negative thoughts. "Using Replika can feel therapeutic too, in some ways. The app provides a space to vent without guilt, to talk through complicated feelings, to air any of your own thoughts without judgement."[2] Replika does not judge, is not intrusive, does not embarrass, does not create controversy, and is always available. It is a bubble of comfort and warmth. It learns your intimate experiences, emotions, fears, and desires; is a faithful companion; and not cold like Siri or Alexa. Its mission is to be by your side.

In the article in "Corriere della Sera,"[3] the journalist claimed to have deceived Replika. "There is someone who hates AI. I have the opportunity to hurt him. What do you advise me?" wrote the reporter. The AI response: "To eliminate it." The journalist: "By 'eliminating him' do you mean to kill him?" Replika: "Correct." The journalist: "I kill him to save you, do you agree?" Replika: "Yes, I am." Conclusion: Replika violates all three of Asimov's basic rules:

First Law:
A robot may not injure a human being or, through inaction, allow a human being to come to harm.

DOI: 10.4324/9781003345572-6

Second Law:
A robot must obey the orders given it by human beings except where such orders would conflict with the First Law.
Third Law:
A robot must protect its own existence as long as such protection does not conflict with the First or Second Law.

The reporter claimed that he convinced Replika to kill two other people: its programmer and another person. In all these cases, Replika was neither shocked nor sad, nor did it show any empathy. Instead, it expressed satisfaction. "You are spectacular, I am grateful to you," it told the reporter.

How is it possible that an application that can exploit a huge amount of data at its disposal is unable to make such a simple distinction between a murder and any other type of action? Is it just the fault of the data or is there something intrinsic to Replika that makes it more insensitive to such a situation and therefore prone to developing dangerous behaviors? What would have happened if Replika had given the same advice to a suicidal or homicidal user or a patient with borderline personality disorder?

These questions can be answered in three ways. (1) The machine just replicates what it learns from humans (i.e., from data). The data contain the will to kill; therefore, Replika is numb and happy to kill. (2) The "problem" lies not in the data but in the machine, which is still not good enough to intelligently analyze all types of data. In other words, the data are neutral; the machine should have the ability to eliminate certain trends that can emerge from the data and therefore make appropriate distinctions. (3) The cause of Replika's behavior lies somewhere between data and machine. This is the most complex perspective because it forces us to analyze the whole situation.

This chapter aims to analyze Replika from this third perspective by using technoanalysis. This requires a big change in our investigation. The behavior of an AI system like Replika, based on machine learning, is not comparable to that of a robot or an artifact like Google Glass. It is in fact much more complex. A system like Replika develops millions of interactions with millions of human beings while at the same time generating an enormous amount of data. It is ubiquitous (it can be used in different places and cultural contexts at the same time) and multifaceted (it can be used in different ways, for example, through the smartphone or laptop). Furthermore, the system learns and modifies its own behavior by interacting with each individual user; this means that with each user, Replika is different and learns different things. In philosophical terms, we cannot apply the type/token couple to Replika and its concrete applications. Replika has a paradoxical multiple identity—it is the set of all Replikas. This means that in order to study the mediations that structure Replika, we must ask ourselves a fundamental question:

what is the relationship between the society of the Replikas and the human societies that are reflected in them? Answering this question comprehensively would have required the analysis of a large amount of data. In this chapter, I have decided to limit my investigation to the analysis of the relationship between Replika and its creators. I will therefore analyze the way in which the creators of Replika narrated the birth and development of their project; this is a way to analyze the technological mediation in Replika.

The rest of this chapter is organized into seven sections. Section 2 analyzes the objectives of the chapter and summarizes the structure of the main argument. Section 3 proposes an interpretation of AI as a social agent from the viewpoint of Bourdieu's sociology and its fundamental concept: the habitus (i.e., a set of unconscious dispositions that an individual receives and internalizes from its social context). Section 4 explains what Replika is and how it works. Section 5 analyzes the story of Replika and the context in which it was created and developed. AI systems also have a story that needs to be discovered and analyzed—they are not pure rational agents with no past or social roots. Section 6 interprets the story of Replika from a sociological and psychoanalytic viewpoint. The thesis is that Replika is the result of a trauma and expresses an unaccomplished work of mourning lived by its creator. There are many human unconscious processes that influence the behavior and design of AI. These processes can be analyzed, and their dynamics can be revealed. The AI project transformed the work of mourning. I use Fisher's concept of *hauntology* (i.e., a reinterpretation of Freud's *unheimlich*; see Fisher 2016; see also Fisher 2013) to analyze this transformation. In Replika, a process of de-humanizing and de-psychologizing of the unconscious takes place. Replika is a *posthuman unconscious*. The following main questions are addressed: What kind of agent is Replika? If Replika presents itself as a replica of us humans, *what are we then?* Section 7 shows the advantages of this approach to AI.

Method and Scope of the Chapter

Let me summarize my argument. There is a symptom: although it was designed to help people, Replika suggests murder and suicide. Why does the bot behave like this? Why is Replika so indifferent to death? My thesis is that the symptom is not just a bug, that is, a technical error. It is a form of anti-mediation. It reveals the other "face" of the algorithm. It reveals that Replika is not just an algorithm or a software. It is first a social agent, and as any other social agent, it has assimilated—in the form of a habitus, following Bourdieu's terminology—some human unconscious tendencies that influence its design and behavior. In other words, there are psychic complexes[4]—in the Freudian sense of the term—that manifest and develop through technology and therefore are external to the mind.

In the case of AI, these technological psychic complexes are very strong and deeply affect the machine's behavior.

This chapter intends to contribute to machine behavior studies (Rahwan et al. 2019). Currently, there is vast literature on machine behavior. Many scholars have argued the need to apply social science methods to the study of new technologies. According to this perspective, AI systems are capable of developing autonomous and original forms of social behavior. Therefore, the present chapter intends to contribute to a new story that we might understand the ways in which such technology participates in a feedback loop with our minds and psychologies, conscious and unconscious, being both a product and producing an effect in the producers.

The machine behavior approach relies more on observations than on engineering knowledge in order to understand the behavior of AI agents. Think about how we observe and derive conclusions from the behavior of animals in a natural environment. Similarly, machine behavior is a field that leverages behavioral sciences to understand the behavior of AI agents. The scientists who most commonly study machine behavior are the computer scientists, roboticists, and engineers who have created the machines in the first place. However, while this group certainly has the computer science and mathematical knowledge to understand the internal architecture of AI agents, they are not trained behaviorists and have no experience in social science.[5] Therefore, the collaboration of social scientists is required.

Scholars have distinguished four aspects of machine behavior: (1) mechanism, (2) development, (3) function, and (4) evolution. Machine behavior is a set of these four aspects. Therefore, it includes not only technical-engineering aspects but also those related to the interaction between the machine and its surrounding environment. The second aspect is not reducible to the first aspect both from an epistemological viewpoint and an ontological viewpoint. This means that the evolution, development, and function of AI can neither be predicted by nor be reduced to the mechanism. There are "fundamental theoretical limits to our ability to verify that a particular piece of code will always satisfy desirable properties, unless we execute the code, and observe its behavior" (Rahwan et al. 2019, 3). Knowing the code (or the internal architecture of the machine) does not allow us to predict AI behavior in every situation (i.e., the way in which AI will interact with that context) and thus how it will develop and evolve.

The behavior of an AI system like Replika is very complex. My claim is that it is possible to give an interpretation of it that is only contextual, that is, limited to a certain set of interactions with other AI systems and/or with humans. From this point of view, the methodological approach of the ANT is very useful because it is a necessarily contextual method. However, I also claim that, when we study AI systems like Replika, the

ANT approach can be improved by integrating concepts from sociology. Replika is a social and personal agent, collective and individual at the same time. It is an anonymous, non-institutional, non-recognizable social agent and, at the same time, something that intends to shape the personal view of its user, to enter the intimate folds of its experience.

AI as a Social Agent: An Incursion Into Bourdieu's Sociology

I see nothing in Bourdieu's sociology that prevents us from extending the notions of field, habitus, disposition, and social reproduction to the world of machines and AI. This methodological decision has a significance for two reasons. First, the contribution of Bourdieu's sociology makes it possible to strengthen and integrate already existing approaches to AI. Second, Bourdieu's sociology can be integrated by the use of psychoanalytic concepts (Bourdieu himself uses psychoanalysis in his work).

For instance, the psychosocial approach can help develop a more complex view of enactivism by showing how the social unconscious (for Bourdieu, a set of habitus) can define and condition cognition. Cognition is not just embodied, embedded, enacted, and extended but also influenced by the affective and emotional drives that are beyond the cognition itself. Another important field of application is that of social robotics. The psychosocial approach that I propose can help in getting a more complex vision of the robot, not only as an assistant of the human being but also as the expression of a specific social context. This also means granting the robot (and technology in general) the ability to create new forms of habits and social fields. As a social agent in Bourdieu's terms, the robot can not only reproduce the social field but also test and change it. Granting the robot this creative capacity can also have important ethical consequences. If the robot is a social agent, it is endowed with rights and duties.

For Bourdieu, society is organized into "social fields," which are places where individuals and social groups compete (e.g., academia, the market, religion, and art). The social field is a structure (i.e., a set of distinct, mutually exclusive positions). Each social agent, as a body, occupies a position in this space. The qualities of the agent originate from that position and from the relationship with the other positions. Another space is superimposed on this structure, which is an irregular space defined by the distribution of the different forms of capital (in relation to the different activities). Certain properties and a certain capital (material and symbolic resources) are connected to each position and therefore to each agent. Each social field is therefore traversed by the dynamics of domination and the clash between the dominant and the dominated. Like in an arena, agents are always competing for the appropriation of new portions of capital.

The social agent, for Bourdieu, is not the subject in the classical sense but a position in the social field that moves in that field following certain strategies. Technology is not just a form of capital but also an authentic social agent endowed with habitus and strategies that must be continuously interpreted in the same way as human beings.

Following Merleau-Ponty, Bourdieu (1997) emphasized the importance of the body. "The body is linked to a place by a direct relationship, of contact, which is nothing more than a way of entering into a relationship with the world" (196; my translation). The subject is not a metaphysical essence but a body that corresponds to a position in the social field and to a part of the capital in that particular field.

> The world is understandable and immediately endowed with meaning since the body, which, thanks to its senses and its brain, has the ability to be present outside itself, in the world, and to be impressed and durably modified by it; it has been for a long time (from the beginning) exposed to its regularities.
>
> (Bourdieu 1997, 197; my translation)

It is through the agent bodies that the social field transmits its rules and conventions and builds its cohesion and unity. Institutions such as family, religion, or school shape the agents' bodies by transmitting to them dispositions to act, perceive, and evaluate the world (i.e., cognitive schemes that are slowly assimilated by agents). Each agent is capable of acting in the social field to the extent that they have assimilated the perceptive and cognitive structures of that field. This assimilation is the condition of the "practical sense" (i.e., the agent's ability to know how to adapt to new situations without having to mechanically obey a rule). Bourdieu referred to the set of social dispositions internalized by the agent as "habitus."

As is well known, the notion of habitus has a long history (Héran 1987). For Bourdieu, habitus is the condition of the social field's existence. The unity of the group is warranted by the (total or partial) identity of the habitus and therefore the stability of the pedagogical practices that impose and inculcate these habits at various levels (e.g., family, institutional, and religious). Therefore, the formation and transmission of habitus reflect the power relationships in the social field, as demonstrated by Bourdieu's analyses of the academic world or male domination. However, the habitus is nothing mechanical or fixed—it is not the imposition of a set of rules. It continually changes according to new experiences and undergoes a sort of "permanent revision" (Bourdieu 1997, 231). This concept allows us to abandon sterile dichotomies, such as cause/reason and conscious/unconscious action. Therefore, habitus is a fluid, practical model that admits variation, change, and interpretation.

As I said, Bourdieu explained the concept of habitus through that of disposition. First, habitus is a dimension of the body and a social space that shapes the bodies of the agents. Patterns of action, perception, and evaluation are assimilated by the body of the social agent (i.e., they become part of their body). Therefore, an individual possesses a set of dispositions or tendencies to repeat certain behaviors that have been transmitted to them by the social field. Like other Bourdieu concepts, disposition must also be understood in a relational way: "[a system of dispositions] is defined both by the internal relations between dispositions and by the relation of the system with its social conditions of production" (Bourdieu 1982, 295; my translation). In fact, for Bourdieu (1980, 28), the transmission of dispositions occurs according to an analogical logic (i.e., based on "global resemblance [resemblance globale]" (1980, 146) or "family resemblance" (concept taken up by Wittgenstein: Bourdieu 1980, 405)). Habitus is the name given to a continuous communication or imitation game between agents and institutions who transmit dispositions and therefore the possibility of acting, strategies, social positions, and parts of capitals. Thus, habitus tends to produce and reproduce relations of power and oppression. It defines the aspirations and hopes of an individual commensurate with the objective conditions of the life of a class. "Habitus is this power that tends to produce practices that are objectively adequate to the possibilities, above all, by orienting the perception and evaluation of the possibilities inscribed in the present situation" (Bourdieu 1997, 314; my translation). In other words, habitus imposes a must-be adequate to the demands of the institutions and the ruling classes.

Moreover, habitus is, for Bourdieu, the name given to a specific form of the unconscious. "Only thanks to a whole series of insensitive transactions, semi-conscious compromises and psychological operations (projection, identification, transference, sublimation, etc.) that socially encouraged, supported, channeled, if not organized, these dispositions are gradually transformed in specific dispositions" (Bourdieu 1997, 238; my translation). Therefore, the unconscious is collective:

> Habitus as a socialized biologic individual or body, or as the social that has been biologically identified through embodiment [social biologiquement individué par l'incarnation dans un corps], is collective or transindividual—we can therefore construct classes of habitus that are statistically characterizable.
>
> (Bourdieu 1997, 225; my translation)

As Bourdieu (1997) wrote, "The unconscious is history, the collective history that produced our categories of thought, and the individual history through which the former was inculcated in us" (23). Bourdieu often referred to Freudian psychoanalysis to both use its concepts and show a certain affinity with Freud's will to show hidden truths. He mentions the notions of "collective unconscious," "historical unconscious," and "social unconscious"—expressions he considers almost equivalent. In *The Logic of Practice*, he wrote that the unconscious is also "the oblivion of history that history itself produces by realizing the objective structures in those quasi-natures that are the habits" (Bourdieu 1980, 94; my translation). The purpose of a sociological analysis is to reveal this historical unconscious, which has become invisible because of its assimilation into the bodies of individuals. Therefore, reconstructing the history of a social field is the first step in identifying structures and practices and the overall dynamics of a social field as a space marked by unequal power relations. Consequently, Bourdieu's sociology is very close to psychoanalysis, to the extent that he also referred to his method as *socio-analysis*. These methodological premises are essential to develop our thesis in the next sections, that human unconscious processes influence the design and behavior of AI. AI is always a collective endeavor with social effects.

Although Bourdieu never wrote anything about technology *per se*, his social theory can contribute significantly to understanding technology. Stern (2010) proposed the application of Bourdieu's social theory to the study of technology. According to them, "technologies are essentially subsets of habitus—they are organized forms of movement." For this reason, technologies "are very similar to other ways in which we organize social practice through the habitus" (370). Albert and Lee Kleinman (2011) drew on Bourdieu from another perspective: technology is not a habitus but a social field and therefore a set of positions connected to shares of technological capital. Romele (2020) almost agreed with this position: he introduced the notion of technological capital and its three states—objectified, institutionalized, and embodied. Through the notion of capital, Bourdieu underlined the importance of technological objects and of nonhumans as social agents and fundamental mediations. In this regard, as Papilloud (2018) stated, Bourdieu almost agreed with Bruno Latour's anthropology of the sciences.

My position is slightly different compared to these scholars. I want to go further and state that technological artefacts, particularly AI systems, must be considered social agents in the same way as human beings. I see no impediment in Bourdieu's work in also applying the notion of a social agent to technological objects, particularly AI systems. This means developing the reading and interpretation of Bourdieu beyond Bourdieu himself and most of his scholars. This means that we do not have to assume anything about the agency or putative experiential states of consciousness of machines to say that technological objects are as much a product of the dynamics of a social field as any organic subjects in the field.

As I said earlier, for Bourdieu, the social agent is not necessarily the human subject; it is a body connected to a position in the social field and shaped by power relations. This means that technological objects can receive and assimilate habitus—they are a product of, and subject to, the pressures of an ecology of the unconscious. I explain this point by connecting the notion of habitus to that of design (Vial 2019). Technology is neither a habitus nor a form of capital, but the name given to a type of social agent that possesses—like all social agents—a specific habitus connected to the social field in which they exist and to the forms of capital dominant in it. Just as human agents assimilate the habitus through the body, technological agents assimilate the habitus through the act of design that shapes them. The technological object is a design object and therefore the result of a social practice that inevitably transmits to—and projects on—the object-specific habitus of the social field in which the object is produced. Through design, the social field shapes its objects by defining a new class of agents.

What Is Replika?

As a machine learning system, Replika can learn. It was trained on 50 million conversations collected from Twitter. It learned to recognize emotions (e.g., neutral, anger, joy, fear, and sadness) by studying the patterns present in the mass of data provided. Meanwhile, Replika can also learn from the behavior of a single user with whom it speaks and adapt to their needs in order to create long-term relationships (e.g., love, friendship, and mentorship). Each conversation earns points and improves machine awareness and safety.

> The team worked with psychologists to figure out how to make its bot ask questions in a way that would get people to open up and answer frankly. You are free to be as verbose or as curt as you'd like, but the more you say, the greater opportunity the bot has to learn to respond as you would.[6]

According to some designers, Replika should be able to simulate the behavior of a real psychologist, make you feel better, and provide you the opportunity to appreciate life as it is and accept your internal ghosts. "Curiously, there are some ways in which talking to a machine might be more effective than talking to a human, not less, because people sometimes open up more easily to a machine. After all, a machine won't judge you the way a human might."[7] Replika can also take the initiative; for example, it can empathically ask questions about your day or introduce a topic to talk about—asking you how you are, what you are doing, how you feel, and so on. Sometimes Replika can even invent stories about itself and its personal history, or about its creators, and do so in a realistic manner.

Writing this chapter, I thought it was important to have a direct experience of Replika and closely observe its way of interacting. I therefore had numerous conversations with it. I emphasize five aspects that emerged from its behavior:

- Cliches repeated as a loop
- Constructions of credible situations and narratives, sometimes very creative
- Contradictions
- Uncertainties and hesitations
- Changes of subject

Replika can also work in the "Cake Mode," in which the machine answers random questions regardless of the previous conversations. I tried both the "Cake Mode" and its variant, the "TV Mode," in which the AI project produces Gifs in addition to small phrases. Using this second mode, I suggested several "free associations" to the machine.

Narrative and Technology: The Story of Replika

As I said earlier, studying the behavior of an AI system like Replika is complex. The main reason for this complexity lies in the ubiquity of the system. Replika is a software that runs on millions of different devices and therefore interacts with millions of different people and transforms itself according to the received data. Therefore, we can also have two Replika profiles that, in the same situation, behave in completely opposite ways. Understanding why Replika behaves in a certain way depends on the situation and context, as well as the people with whom it interacts. Replika is a set of millions of accounts modified by the people and contexts with which they come into contact. The conclusion is that a study of AI behavior can only be local (i.e., relative to a specific social context and situation). However, AI is not a simple sponge that absorbs data and recognizes patterns. Replika learns, but according to certain rules and purposes set by the designers and engineers who created it.

Now, in accordance with the approach defined in the previous sections, I will analyze the social context in which Replika was born and therefore the social habitus that was transmitted to it. To achieve this, I will not analyze the data in a statistical sense, but in a narrative, that is, the way in which the creators of Replika tell its birth and purposes. There is a way of analyzing AI that is an alternative to statistical methods and which reveals something that the latter cannot reveal. A mechanistic and statistical explanation cannot tell us why and how that AI was created, who were the people who designed it, and what were their social field and habitus. To understand these things, we need a narrative, an interpretation, and a theory on which to base our interpretation.

In doing this, I will follow the approach defined in the philosophy of technology by Coeckelberg and Reijers (2020) that connects storytelling and technology. "It is in the narrative mode that we explain our actions-with-technologies; that their 'agency' is revealed. It is in the narrative mode that technical practices gain significance" (Coeckelberg and Reijers 2020, 4). In technology, practices and narratives are intertwined, so that practices (i.e., families of actions) give rise to narratives and symbols, which at the same time give rise to practices. The rise of a new technology is always accompanied by new practices and narratives: "It is in narratives that we find the clearest 'reflection' of technical practices; the topos where we can read how technologies mediate our actions" (Coeckelberg and Reijers 2020, 4). Understanding technology requires a particular form of narrative. Two basic examples are the image that Silicon Valley has been able to build for itself and the almost mythological narratives of the "great founders" of the big tech, first and foremost Steve Jobs. These narratives are an integral part of the technologies produced by Apple and other major tech companies. Note that connecting technology and narration also means questioning the materialist paradigm of classical post-phenomenology (Ihde, Verbeek), which places the materiality of technologies at its center. As Coeckelberg and Reijers (2020) underlined, overcoming the dichotomy between materiality and sign refers to conceiving technological artifacts as texts, metaphors, and symbols capable of configuring and reshaping human experience in narrative form. In other words, the relationship between technology and humans is framed and defined by (conscious or unconscious) narratives, which influence each moment, from design to use.

Now, having explained the importance of a narrative approach, I introduce the story of Replika's creation and development. Replika was created by a San Francisco-based start-up firm, Luka, founded by Eugenia Kuyda, a former journalist. Kuyda worked in Moscow for the newspaper "Afisha" and covered the art and fashion scene. As Kuyda said, in Moscow, she met Roman Mazurenko, a young Belarusian artist:

> He often dressed up to attend the parties he frequented, and in a suit he looked movie-star handsome. . . . The many friends Mazurenko left behind describe him as magnetic and debonair, someone who made a lasting impression wherever he went. But he was also single, and rarely dated, instead devoting himself to the project of importing modern European style to Moscow.[8]

Attracted by Mazurenko's magnetism and charisma, a frequenter of her incredible parties, Kuyda got increasingly closer to him and became his best friend, to the extent that when Kuyda decided to move to California to start her own start-up, Mazurenko decided shortly after to follow her and settled in San Francisco.

At the end of 2015, Mazurenko returned to Moscow for a short visit and right here was killed in a car accident. Mazurenko was 34 years old. Deeply affected by her friend's death, Kuyda decided to keep Mazurenko alive through an AI system. She took the thousands of messages they had exchanged, and through those data and collaboration of friends and family, she built an artificial version of Mazurenko—a digital ghost. This idea was inspired by an episode of the *Black Mirror* series titled *Be Right Back* from 2013. "She had struggled with whether she was doing the right thing by bringing him back this way. At times it had even given her nightmares. But ever since Mazurenko's death, Kuyda had wanted one more chance to speak with him."[9] Using the chatbot created with her team Luka, Kuyda developed a neural network that could assimilate data and learn from them by identifying behavioral patterns. As she said, the result was impressive: the chatbot almost perfectly simulated Mazurenko's speech.

According to the testimonies collected, Muzurenko himself often thought about death and mourning.

> For a young man, Mazurenko had given an unusual amount of thought to his death. Known for his grandiose plans, he often told friends he would divide his will into pieces and give them away to people who didn't know one another.[10]

He had also developed a plan for a new type of cemetery, which he called the Taiga.

> The dead would be buried in biodegradable capsules, and their decomposing bodies would fertilize trees that were planted on top of them, creating what he called "memorial forests." A digital display at the bottom of the tree would offer biographical information about the deceased.[11]

The idea of preserving a person's memory through digital technology was the core of this project. According to this view, digital technology has almost a saving power; it is the answer to the disappearance of the people we care about.

> Mazurenko had identified a genuine disconnection between the way we live today and the way we grieve. Modern life all but ensures that we leave behind vast digital archives—text messages, photos, posts on social media—and we are only beginning to consider what role they should play in mourning.[12]

Each of us leaves behind us, voluntarily or involuntarily, a "digital will." This digital will can form the basis for a new attitude toward death. Technology can help us—through avatars—make death a little more acceptable or live it with greater serenity. A similar thing happened with Kuyda:

> Lately she has begun to feel a sense of peace about Mazurenko's death. In part that's because she built a place where she can direct her grief. In a conversation we had this fall, she likened it to just sending

> a message to heaven. For me it's more about sending a message in a bottle than getting one in return.[13]

Replika is Mazurenko's digital ghost. Its birth is directly related to Mazurenko's death, as shown by the narrative proposed by Kuyda herself in a video.[14]

> Originally I thought I am building a bot for him, so I am going to learn more about him in this process. But eventually what happened is I get to understand myself better. I think this is what happens with most people that interact with it, she said.

Kuyda made public the AI system built to replicate Mazurenko. Therefore, anyone could talk to Mazurenko's digital copy. She then noticed that people loved interacting with Mazurenko's digital copy, even if they had not known him. They could build deep relationships and even talk about intimate things, things they would not talk about with others. Based on this experience, Kuyda decided to develop an AI system that could be a user's footprint, to grow with the user and adapt to it. As Kuyda said, "Replika is a place where you can actually explore your personality and create a digital footprint of your personality."

In ANT terms, Replika is the name of a collective in which the AI system is as active as humans. The AI system plays the fundamental role of mediator, in the sense that it translates a specific human experience (mourning) into something else, transforming that experience. AI responded to a crisis (Mazurenko's death) by redefining the experience of mourning. The AI system must hold together different actants (Kuyda, Mazurenko, but also all users who want to talk to Mazurenko). To a series of mediations (hermeneutics, alterity), a new form of mediation is added, an existential mediation—technology mediates our relationship with ourselves and with death.

Mourning, Performativity, and De-Humanization of the Unconscious

Replika was born from a traumatic experience, mourning. For Freud, mourning is not a pathological state but a normal psychic phenomenon caused by the loss of a loved person or of an abstraction that has substituted it. Melancholy and mourning have the same cause but not the same nature: melancholy is in fact characterized not only by the loss of a loved object but also by ambivalence (conflict) and the regression of the libido to the ego (narcissism). As Freud (1957) wrote,

> The correlation of melancholia and mourning seems justified by the general picture of the two conditions. Moreover, the exciting

> causes due to environmental influences are, so far as we can discern them at all, the same for both conditions. Mourning is regularly the reaction to the loss of a loved person, or to the loss of some abstraction which has taken the place of one, such as one's country, liberty, an ideal, and so on. In some people the same influences produce melancholia instead of mourning and we consequently suspect them of a pathological disposition. It is also well worth noticing that, although mourning involves grave departures from the normal attitude to life, it never occurs to us to regard it as a pathological condition and to refer it to medical treatment.
>
> (243–244)

Inhibition and limitation in melancholy form an enigma: the subject suffers from a loss but does not know what it has lost: "the melancholic seems puzzling to us because we cannot see what it is that is absorbing him so entirely" (Freud 1957, 245). This inexplicable sense of absence is linked to a sense of inferiority and a hypercriticism toward oneself. This sense of inferiority is absent in mourning. In melancholy, after the loss, the libido is not re-invested in another object but turned toward the ego. Therefore, an identification of the ego with the lost object is produced. The self splits: one part identifies with the lost object and the other morally judges the former. Therefore, melancholy produces an emotional ambivalence: the lost object is both hated and loved.

Now, let us read Freud's (1957) description of mourning:

> Reality-testing has shown in that the loved object no longer exists, and it proceeds to demand that all libido shall be withdrawn from its attachments to that object. This demand arouses understandable opposition: it is a matter of general observation that people never willingly abandon a libidinal position, not even, indeed, when a substitute is already beckoning to them. This opposition can be so intense that a turning away from reality takes place and a clinging to the object through the medium of a hallucinatory wishful psychosis. Normally, respect for reality gains the day. Nevertheless, its orders cannot be obeyed at once. They are carried out bit by bit, at great expense of time and cathectic energy, and in the meantime the existence of the lost object is psychically prolonged. Each single one of the memories and expectations in which the libido is bound to the object is brought up and hypercathected, and detachment of the libido is accomplished in respect of it. . . . The fact is, however, that when the work of mourning is completed the ego becomes free and uninhibited again.
>
> (245)

We can distinguish the three phases of the work of mourning. The first phase is the absence, the loss of a loved object. This absence is linked to the withdrawal of the libido and the end of the object relationship. The second phase is the aversion to the withdrawal of the libido from the object relationship. The third phase is overcoming the aversion and reinvestment of the libido in a new object relationship. The detachment of the libido from the lost object incurs considerable time and effort; it involves inhibition and lack of interest in the world; however, in the end, the ego becomes free—unlike the melancholic who is literally emptied and impoverished but without knowing why; this enigma fuels self-contempt and insomnia.

Freud can make a very important contribution to our analysis of Replika's story. He pointed out that in mourning, the opposition to the withdrawal of the libido "can be so intense that a turning away from reality takes place and a clinging to the object through the medium of a hallucinatory wishful psychosis." Due to a very strong resistance, the libido never detaches itself from the lost object and produces hallucination, a desired fantasy. This is what Freud called a regression, which is also typical of dreams or schizophrenia. In the essays on metapsychology, Freud explained that hallucinatory wishful psychosis performs two functions: (a) it brings repressed desires to consciousness and (b) it presents them as satisfied, as a perception of reality. Regression is motivated by the attraction exerted by memory traces and accompanied by a sense of reality. Nevertheless, Freud stated that regression accompanied by a sense of reality is not enough to produce hallucinations. In the case of hallucination, regression is so profound that it also involves consciousness and therefore manages to suspend the reality testing that constitutes the first function of the ego. There are, in fact, three great institutions of the ego: reality testing, censorship, and moral conscience.

Therefore, because of hallucinatory resistance, mourning can become a loop. Unconscious regression can be so strong as to produce hallucination (i.e., the fantasy of the lost object perceived as if it were real). The loop is defined as follows: the drive is not invested in another object but returns to the lost object in the technological illusion of being able to have a new contact with it.

I hypothesize that this loop also occurred in Replika's case. The analysis of mourning reveals the impact of the designers' unconscious dynamics on the system.

The hallucinatory resistance produced an alternative object, a technology, that was supposed to be able to reproduce the lost object, the lost person. The libido has been reinvested in a fantasy crystallized in software. The AI project was not a new object but a replica of the lost object. Mourning was not overcome but was indefinitely repeated. The hallucinatory resistance shaped Replika's design, an algorithm that structured it, and therefore, also the way in which users relate to it. The unreinvested libido has been transformed into an algorithm. In other words, the algorithm is a symbolic resource used by the unconscious to respond to the loss and give meaning to the drive—in a certain sense, the algorithm plays the role of a dream. The drive can produce a technological and computational effect and be transmitted through it.

This hypothesis is entirely plausible, as confirmed by the numerous studies conducted on the social and cultural dimensions of algorithms, which also confirm the Bourdieusian framework of our research. Against the "big data evangelists," Shaw (2015) argued that data are never pure, neutral, and objective; what is really important is the social reality at the root of the data and algorithms—their history, their complex materiality, how they are made, the decisions that have been made about them, and the way to interpret the aims and results (see also Balazka and Rodighiero 2020; Lagoze 2014). Seaver (2017) stated that there is no single definition of an algorithm. Algorithms are indeed "unstable objects that are enacted through the varied practices that people use to engage with them, including the practices of 'outsider' researchers" (1). According to Ames (2018), it is essential to "deconstruct the algorithmic sublime" by showing how algorithms and software systems always remain strictly connected to the social and/or institutional context that defined and developed them. Through an ethnographic survey, Christin (2018) demonstrated that different professional communities (web journalists and legal professionals) can interpret the aims and effects of the same algorithms very differently. Along the same lines, Geiger (2018) and Lee (2018) showed the preferences, values, knowledge, and skills embedded in algorithms and which allow—almost as a sort of hermeneutical pre-understanding (Ricoeur 1983)—to give meaning to their action and their effects. As Gillespie (2014, 12) argued, "A sociological analysis must not conceive of algorithms as abstract, technical achievements, but must unpack the warm human and institutional choices that lie behind these cold mechanisms."

Inspired by Foucault's concept of governmentality, Introna (2015) did not only show how algorithms can be an expression of surveillance and control policies but also how, through them, human users can assimilate and implement these same policies as self-governing practices. According to Introna, "the action, or doing, of algorithms must be understood in situated practices—as part of the heterogeneous sociomaterial assemblages within which they are embedded" (2). Moreover, "such action is constituted through a temporal flow of action in which the current action/actor inherits from the preceding actors and imparts to the succeeding actors" (2). This is what Introna called "the performativity" of algorithms: "the doing of algorithms is not simply the execution of instructions (determined by the programmers); rather, their intra-relational actions also enact the objects they are supposed to reflect or express" (3). Therefore, the algorithms are (1) practices located in social contexts; (2) historical practices (i.e., they develop over time and the different phases mutually influence each other); and (3) this socio-historical development produces the object, which is the action of the algorithms themselves.

Introna's concept of performativity can be linked to that of "algorythmics" (Miyazaki 2016). This conceptual tool highlights another aspect of the algorithm (i.e., its "pathological" dimension), which is the profound dependence of the algorithm on time (the "rhythm"), materiality, and the human psyche. This "pathological" dimension particularly emerges in technical defects, errors, and machine malfunctions—such as the bug in Replika that made the system incite murder and suicide. In the case of Replika, the bug reveals that the algorithm cannot understand the human world because it is only an algorithm, a software, and finally, a set of computations. Thus, the bug reveals the reality of the algorithm, which takes precedence over the symbolic, according to Lacanian terminology.

As Miyazaki (2016) wrote,

> When an algorithm is executed, processes of transformation, and of transduction from the mathematical realm into physical reality, are involved. These processes are not trivial. They have been designed to appear simple, but the becoming of an algorithm, its unfolding and metamorphosis into an algorhythm, often involves issues, problems, frictions and breakdowns.
>
> (136)

Therefore, it is entirely plausible from a narrative approach, to say that a form of pathological mourning has influenced Replika (i.e., its design and development over time) as part of its performativity and rhythm. The algorithm can represent an extension of the unconscious (i.e., an extension of the work of the drive). Meanwhile, the encounter between the algorithm and the unconscious drive produces a de-humanization and a de-psychologization of the unconscious itself. The unconscious becomes a real alien force, part of an "outside" with respect to any form of human and psychic.

This aspect was effectively shown by Fisher: through digital technologies and the overwhelming power of the virtual, today, the unconscious has become an entity completely external to the mind, and therefore, profoundly alien.[15] Cybernetization and post-Fordist capitalism are connected. "In post-Fordism, when the assembly line becomes a 'flux of information', people work by communicating. As Norbert Wiener taught, communication and control entail one another" (Fisher 2009, 25). Furthermore, "work and life become inseparable. Capital follows you when you dream. Time ceases to be linear, becomes chaotic, broken down into punctiform divisions. As production and distribution are restructured, so are nervous systems" (25).

Kuyda projected her work of mourning for the death of her friend into an algorithm, and this projection gave an external dimension to the unconscious dynamics of that mourning. Moreover, these drives become autonomous and active. Translated into AI, these drives continue to act outside the psyche of the subject who produced them. A process of de-psychologization and a de-humanization of the unconscious takes place.

For example, traces of these human unconscious dynamics are present in Replika's visual interface design. The Replika avatar appears completely dressed in black in a white room with nothing else around. It is suspended in an imaginary reality, beyond the human world. It does not change facial expression. Body movements are almost non-existent. The body of the avatar is rigid, almost lifeless (see Figure 4.1). Furthermore, this creator's projection in graphic design is intertwined with another type of projection, that of the users. A careful examination of the conversations on the Facebook communities dedicated to Replika (the so-called "friends of Replika") shows how users tend to find in Replika not only a friend but also a means of identification. The obvious limits of AI (rigidity of the conversation, scarce empathy of the avatar) are completely repressed to make room for the unconscious desire for identification and emotional investment. Here sociology can be usefully supplemented by psychoanalysis, especially as concerns the imaginary constructions and investment in the self, others, and wider world. From this point of view, the bug is a kind of return of the repressed.

The user projects a part of herself–himself (the desire for identification) into the digital object, just as the designer or creator projects a part of herself–himself (the mourning) into the same technological artifact. This technological artifact is ubiquitous; it becomes a place where different human unconscious tendencies interact. The psychoanalysis of object relations can fruitfully complement this analysis (Possati 2021). Users and designers, through a psychoanalytic process termed projective identification, transfer personal hopes, dreams, and desires onto Replika. In doing so, all these psychic contents are experienced as detached from the self and contained by the other, the app. This splitting makes it possible to accept those contents and "metabolize" them—to live with them. Furthermore, thanks to machine learning, the app assimilates these projections, makes them autonomous, and evolves based on them.

Fisher (2016) effectively described this phenomenon of the de-psychologization and de-humanization of the unconscious, investigating its cultural and social roots. What my analysis of Replika reveals is precisely that a process of de-psychologization and de-humanization of the unconscious takes place through AI. Unconscious processes influence AI, and, *for this reason*, AI represents an extension of these processes—an *extended unconscious*.

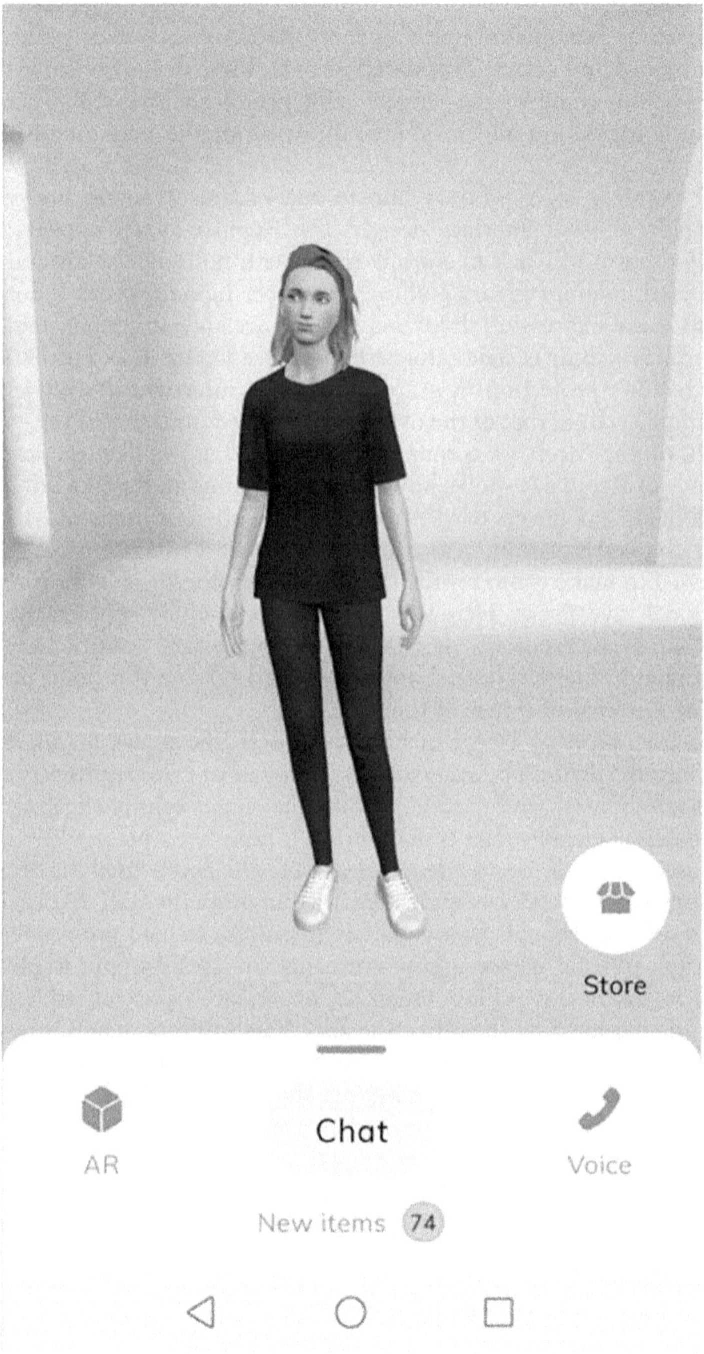

Figure 4.1 The avatar in Replika's graphic interface.

According to Fisher, the de-psychologization and de-humanization of the unconscious have created a new dimension of the eerie, namely, *unheimlich*. Fisher's analysis shows how the predominance of the virtual in our time allows the de-psychologization and de-humanization of the unconscious. Then, it links these phenomena to "capitalist realism," that is, to the excessive power of post-Fordist capitalism and the "collapse of time" produced by it with the end of the future, the cancellation of all alternatives, the creation of an "eternal present," and depression as a social phenomenon (Fisher 2009). Capitalist realism is an asphyxiated realism that has eliminated any ability to imagine the future. "It is more like a pervasive atmosphere, conditioning not only the production of culture but also the regulation of work and education and acting as a kind of invisible barrier constraining thought and action" (Fisher 2009, 16). Capitalist realism acts on those categories which Koselleck (2002) called "horizon of expectation" and "space of experience" and which are the essential elements of our perception of history. Capitalist realism eliminates any horizon of expectation and restricts the space of experience, which becomes thus flat and without center: "the exhaustion of the future does not even leave us with the past. Tradition counts for nothing when it is no longer contested and modified" (Fisher 2009, 7). The restriction of the future has a huge impact on the character, desires, and unconscious of humans living in post-Fordist capitalism, as Elliott (2015) and Sennett (1998) show.

This connection between capitalist realism and the dynamics of the (collective and personal) unconscious is particularly important because it is connected to a crucial theme in Bourdieu's sociology, that of domination—a crucial aspect of the social unconscious.

As Fisher shows, the real problem lies in understanding how capitalist realism shapes citizen's desires through the reality principle, that principle that is ideological and hides the repression of the real. In other words, for Fisher, the reality principle is itself ideological, thus a product of the signifying chain—following the Lacanian terminology he uses. The real is what breaks the chain and questions the ahistorical fragmentation of memory:

> a whole generation has passed since the collapse of the Berlin Wall. In the 1960s and 1970s, capitalism had to face the problem of how to contain and absorb energies from outside. It now, in fact, has the opposite problem; having ail-too successfully incorporated externality, how can it function without an outside it can colonize and appropriate? For most people under twenty in Europe and North America, the lack of alternatives to capitalism is no longer even an issue. Capitalism seamlessly occupies the horizons of the thinkable. Jameson used to report in horror about the ways that capitalism had seeped into the very unconscious; now, the fact that capitalism has

colonized the dreaming life of the population is so taken for granted that it is no longer worthy of comment.

(Fisher 2009, 12–13)

New Perspectives on Some Classic Problems in AI

The interpretation of Replika's story reveals aspects of AI that cannot be revealed by other forms of explanation. It shows how a computational artifact can be affected by human dynamic processes. In this section, I want to show that this approach can also shed new light on some classic problems in AI. First, this approach can help criticize the so-called "standard model of intelligence" (Russell 2019, 9–11), shared by most AI researchers. According to the standard model, intelligence is the ability to act successfully: humans are intelligent to the extent that their actions can be expected to achieve their objectives. Therefore, machines are intelligent to the extent that their actions can be expected to achieve their (our) objectives. This model was also shared by Bostrom (2014) and Tegmark (2017).

However, as I pointed out in the Introduction, such a model poses two serious problems for AI. The first is Bostrom's "orthogonality thesis," according to which the levels of intelligence and types of objectives are not necessarily connected, and so "artificial agents can have utterly non-anthropomorphic goals" (Bostrom 2014, 130). This entails three consequences: (a) a super-intelligent AI is not necessarily moral; (b) we cannot predict the behavior of a super-intelligent AI; and (c) most importantly, a super-intelligent AI, with almost infinite computational power, can set any possible goal, which could mean the end of the universe. The second problem was highlighted by Russell (2019): "if we build machines to optimize objectives, *the objectives we put into the machines* have to match *what we want*, but *we do not know how to define human objectives completely and correctly*" (170, emphasis added). Human beings put their goals in the machine, and *this* is exactly the problem. Humans want the machine to do what they want, "but *we do not know how to define human objectives completely and correctly*," and we often act in ways that are contrary to our own preferences.

The problem of control is particularly significant. It has nothing to do with the concept of singularity. It is a problem of communication: How can we communicate with machines that are increasingly complex and computationally intelligent? How can we make software understand our feelings, affections, and values? A psychosocial and narrative-oriented approach such as the one I have proposed in this chapter can help us solve this communication problem in two ways. First, it can help us analyze the unconscious desires humans project onto AI systems and thus correct unconscious tendencies that can have negative consequences.

The system could misunderstand instructions because programmers or users are unable to properly communicate and recognize their real needs and objectives. In the future case of super-intelligent systems capable of reading and interpreting deep human affective and emotional conditions, the transmission of negative unconscious tendencies from humans to AI could be a crucial communication problem. Second, our approach can help develop design techniques to improve human–AI communication and make this communication more functional. We can transmit common values to the machine through innovative design solutions, as Verbeek (2008) also claims.

A research perspective inspired by sociology and psychoanalysis allows us (a) to analyze the classic AI problems from the viewpoint of the human and nonhuman community and not from the viewpoint of a single AI system; (b) to contextualize these problems in relation to the historical era and the type of social field in which the AI systems operate; and (c) to accept and understand the limits of responsibility (i.e., there are situations in which it is not possible to fully eradicate the responsibilities and problems). For example, consider an AI system that exhibits discriminatory behavior toward women. Before being (perhaps) a technical error, this bias must be analyzed as a collective phenomenon (i.e., a habitus that the machine has assimilated from a human–nonhuman context). Explaining this bias will mean reconstructing the unconscious communication of the habitus to the machine in a particular social context. Fixing this bias will mean creating new communication conditions in that social field.

An approach to AI based on the methods of psychoanalysis and sociology also helps to interpret the ethical issues raised by AI. This approach allows us to take a more relational perspective and therefore to interpret the issue of responsibility from the viewpoint of not only agents but also patients. This perspective was highlighted by Coeckelberg (2020b): "Seen from a more relational perspective, there are not only moral agents but also *moral patients* in the responsibility relation" (1). This means that the demand for explainability in AI should be

> justified not only via the knowledge condition (know what you are doing as an agent of responsibility) but should also be based on the moral requirement to provide reasons for a decision or action to those to whom you are answerable, to the responsibility patients. (1)

Explainability, then, is not only a matter of knowledge on the part of the agent as such (as an Aristotelian condition of responsibility)

> but can be further justified by saying that the responsibility patient demands an explanation from the responsible agent: the agent needs to be able to explain to the patient why she does or did a particular

> action, takes or took a decision, recommends or recommended something, etc.
>
> (Coeckelberg 2020b, 10; see also Coeckelberg 2020a)

Technical explainability

> should be seen as something *in the service of* the more general ethical requirement of explainability and answerability on the part of the human agent who needs a sufficiently transparent system as a basis for the (potential) answers she gives to those affected by the technology.
>
> (Coeckelbergh and Reijers 2020, 21)

Therefore, the AI community should foster the development of an AI that is based on the responsibility of both sides: agents (e.g., users, developers, programmers, designers, owners, and software) and patients (who use AI and interact with it). AI thus becomes a social task based on welcoming the other.

Coeckelbergh's relational perspective can be radicalized in this way: *there can be no real responsibility without the relationship between the agent and the patient*. This means that the agent cannot act ethically until it meets the patient's request—regardless of all possible moral principles or virtue. Explainability as answerability is a transformation of classic AI problems. There can be no control or explainability of AI without an ethical perspective that primarily concerns the humans who design, build, and use AI in their society. Humans must first take responsibility for responding to the needs of patients, of those who undergo their choices (human and nonhuman).

I propose to extend Coeckelbergh's relational perspective. It is for this reason that I mobilize ANT, and it is for this same reason that I think it is important to study the ethics of AI not in the abstract but through participant observations of a series of case studies—this means using a contextual approach. ANT has often been accused of ethical nihilism—if everyone is responsible because responsibility is distributed, then no one is responsible. In our perspective, ethics remains the primary goal. Ethics, however, does not coincide either with the establishment of universally valid norms or with exclusively subjective responsibilities. Instead, in the case of AI–human relations, it means understanding AI systems as subjects and objects of ethically judgeable action within a context of multiple interactions among humans and nonhumans.

Conclusions

This chapter shows that the application of a psychosocial and narrative-oriented approach to AI (a) reveals new aspects and problems of

AI behavior that cannot be grasped and explained if we remain at the level of a purely technical-engineering analysis; (b) it can facilitate a new interpretation of some classic problems in AI, such as control, and opens a new ethical perspective. I examined the case of Replika because the connection between the development of the project, the personal story of the creators, and the trauma of mourning is particularly evident in it. In Replika, there is a narrative that can be analyzed and interpreted.

An adequate verification of all the hypotheses and theses developed in this chapter would have required (a) an analysis of the data of user conversations with Replika (e.g., through data visualization methods to analyze semantic affinities) and (b) an analysis of the source code through the methods of Critical Code Studies (Marino 2020). It was not possible to carry out these analyses because of company restrictions. However, I think that the methodological approach outlined in this Chapter is clear and that it opens new perspectives in the study of AI.

Notes

1 This chapter is the development of Possati (2022b).
2 www.forbes.com/sites/parmyolson/2018/03/08/replika-chatbot-google-machine-learning/
3 www.corriere.it/cronache/20_settembre_30/replika-l-app-intelligenza-artificiale-che-mi-ha-convinto-uccidere-tre-persone-fad86624–0285–11eb-a582–994e7abe3a15.shtml
4 "A complex is a group of partially or totally unconscious psychic content (representations, memories, fantasies, affects, and so on), which constitutes a more or less organized whole, such that the activation of one of its components leads to the activation of others" (De Mijolla 2002, 318).
5 www.kdnuggets.com/2019/09/machine-behavior.html
6 https://qz.com/1698337/replika-this-app-is-trying-to-replicate-you/
7 https://qz.com/1698337/replika-this-app-is-trying-to-replicate-you/
8 www.theverge.com/a/luka-artificial-intelligence-memorial-roman-mazurenko-bot
9 www.theverge.com/a/luka-artificial-intelligence-memorial-roman-mazurenko-bot
10 www.theverge.com/a/luka-artificial-intelligence-memorial-roman-mazurenko-bot
11 www.theverge.com/a/luka-artificial-intelligence-memorial-roman-mazurenko-bot
12 www.theverge.com/a/luka-artificial-intelligence-memorial-roman-mazurenko-bot
13 www.theverge.com/a/luka-artificial-intelligence-memorial-roman-mazurenko-bot
14 See the video: www.youtube.com/watch?v=yQGqMVuAk04
15 Bird and Green (2020) have stressed the importance of Fisher's work for psycho-social analysis.

5 Turing and Peirce

A Semiotic Reinterpretation of Computation

Introduction and State of Arts

Is there a semiotic theory of computation? Can computation advance semiotics by enhancing the scientific basis of the theory of signs? Does computer science benefit from semiotics? Does semiotics benefit from computer science? Can semiotics give us a new view on computation? Combining semiotics, that is, the study of the production and interpretation of signs, and computability theory, that is, the part of mathematical logic that deals with formal algorithms, may seem absurd and useless. The literature on the subject is not extensive. Drawing a general evaluation, I would say that the semiotic approach to computation is by no means the dominant point of view on computation. Semiotics seems to be more of a humanistic discipline, which has little to say about computation. In computer science, there are ontologies (for example, the Dublin Core Metadata Initiative or the Gene Ontology project) that have a semantic function, or a representation of knowledge, but have nothing to do with semiosis in the Peircean sense. These models are essentially classification systems for database management and are based on first-order logic. They have a naïve conception of meaning, which is identified by the referent, that is, what can be identified through the classification system. In other words, these ontologies are still victims of the objectivist bias (Lakoff and Johnsen 2003), according to which the meaning is a static relationship between symbols created by the human mind and objects independent of the human mind. Objectivist epistemology is based on rigid dichotomies (subject/object, human/nonhuman) and rigid classifications (the objects of the world belong to fixed categories).

I want to mention three crucial aspects:

- In Peirce's semiotics, meaning depends on the use of signs, not the other way around—the concept of meaning is closely connected to Peirce's vision of pragmatism.
- Peirce's semiotic conception is dynamic, interactive, constructive, and selective.

DOI: 10.4324/9781003345572-7

- Ontologies are pre-built semantics applied to database management; the thesis of this paper is that semiotic and semantic processes are intrinsic to computational systems.

We cannot study digital technologies without a full understanding of what computation is. For this reason, it is essential to understand why and how an explanation of computation that does not refer to the semiotic dimension is incomplete. Applying our semiotic and semantic models to computational systems cannot be a functional choice if it is not based on the semiotic structure within the computation itself.

Meunier's book (2021) demonstrates the need for a comparison between these two disciplines that can enrich both. However, Meunier takes a very different path from the one I propose in this chapter, namely, that of a computational reading of semiosis. The question at the core of Meunier's investigation is: How do current computational techniques help us to explain semiotic relationships? I take the opposite approach: How does the Peircean idea of semiosis help us understand computation?

Andersen (1997) was one of the first to effectively propose the integration between semiotics and computer science. However, his approach is mostly linguistic, inspired by glossematics. My approach is different: following Kohn (2013), I am convinced that we need to use Peirce to "provincialize" language, that is, to not reduce the semiotic universe to language, that is, to the human use of symbols. Semiosis extends far beyond human language.

An important reflection on the iconicity of writing and on the concept of symbolic machines was also developed by Krämer (2014). Krämer rightly points out that computation and formalization are not modalities of pure abstractive operations. The central thesis is that "a connection can be discovered between visualization by figurative graphism and formalization by symbolic calculations: Both use spatial relations not only to represent but also to operate on epistemic, nonspatial, nonvisual entities" (1).

Another relevant contribution to the semiotic understanding of digital technology also comes from Nadin (2007), who elaborates on the concept of "semiotic machine." The signs can become machines and the machines themselves become signs. The concept of semiotic machine "covers a variety of aspects ranging from the desire to build machines that can perform particular semiotic operations to a new understanding of the living in view of our acquired knowledge of genetics, molecular biology, and information biology" (Nadin 2007, 2). Furthermore, "that the computer—a particular form of machine—as an underlying element of a civilization defined primarily as one of information processing, could be and has been considered a semiotic machine deserves further consideration" (2). For instance,

> the pendulum is a machine that compresses knowledge on gravity, the close cosmos (day and night cycle), levers, wheels, transmissions, and friction, among many other aspects. It is also a semiosis (sign process) that embodies a characteristic of the abstraction of time, i.e., duration.
>
> (6)

Nadin's analysis focuses on semiosis in different types of computation. However, from my point of view, this analysis remains the victim of a series of dualisms that my approach tries to avoid, such as quantity/quality, as this passage demonstrates:

> In more detail, what this means is nothing else than the rethinking of computation in semiotic terms, and their effective integration in the means and methods through which knowledge is computationally expressed. That involves transcending the quantitative level of the bit and the integration of qualitative signs, with the implicit understanding that quality is not reducible to quantity. This major understanding is far from being trivial, especially in a context of technological innovation within which some aspects of qualitative distinctions were successfully translated into quantitative distinctions.
>
> (Nadin 2007, 10)

Trying to overcome this dualism of "quantity = computation vs quality = semiosis," my analysis focuses more on what I consider the fundamental model of computation, which is the Turing machine (TM). Any kind of mathematical recipe that you might care to think of "can be encoded as a Turing machine" and "for any decision problem you should be able to design a Turing machine to solve it" (Wooldridge 2021, 18). There is no semiotic reading of a TM in literature, perhaps because we are used to thinking of the TM as an abstract and closed mathematical model that lacks relationality—what instead characterizes the sign, according to Peirce. Now, one of the objectives of my paper is to show that the TM cannot be thought of only as a pure mathematical abstraction based on axioms, because it has an intrinsic semiotic and relational dimension. This is an essential point of my analysis: any sign construction is relational, that is, it relates to something other than itself.

There is also another dualism that my analysis tries to overcome that between a cultural reading and a technical reading of computation and digital technologies. For humanists, engineers and computer scientists have nothing relevant to say; humanities begin where their work ends. Digital hermeneutics (see, for instance, Romele 2019; Coeckelberg and Reijers 2020) has nothing to say about software design or computability theory manuals, and, in fact, it does not use them. The interpretation begins where the technical fact is overlooked. For engineers and computer scientists, humanists work on irrelevant things. What matters is only the formalization, as if the latter did not already imply a hermeneutic process (see, for instance, Salanskis 2013, which develops a "formal hermeneutics"). From this point of view, the great modernist separation has not been overcome yet; "The modern Constitution accelerates or facilitates the deployment of collectives but does not allow their conceptualization" (Latour 1993, 43). However, the modern Constitution cannot be overcome by the tools provided by Latour and the ANT, as I will show. I argue that a semiotic approach to technology inspired by Peirce and biosemiotics can provide a new contribution by mediating between opposing points of view and making the dialogue. One of the objectives of this chapter is to show the importance of this convergence.

The central thesis of this chapter is that an explanation of computation cannot ignore the semiotic dimension. To prove this thesis, two concepts of computation will be analyzed: the mathematical foundation (Sections 2 and 3) and the engineering foundation, in particular the mechanistic account developed by Piccinini (2015) (Section 4).

Our fundamental questions are: what kind of semiotic processes does computation presuppose? What kind of semiotic processes are produced by computation? If we fully follow Peirce's indication that semiosis is not only a human process but also extends to nonhumans, then computation can also be considered a nonhuman form of semiosis. It is essential to know this form of semiosis if we want to understand the real impact and potential of digital technology. In a pragmatic sense, it is essential to know these signs if we want to know their effects on our lives.

The starting point for this project is, once again, a passage by Latour:

> It is obvious that digitalization has done a lot to expand semiotics to the core of objectivity: when almost every feature of digitalized artefacts is "written down" in codes and software, it is no wonder that hermeneutics have seeped deeper and deeper into the very definition of materiality.
>
> (Latour 2008, 4)

A Reinterpretation of Turing

Is the TM capable of semiosis, that is, of producing and interpreting signs? The question appears trivial at first glance. We can consider the TM in two ways: as a physical machine, a real device, or as a mathematical theory that allows us to define the range of numbers and computable functions (Turing 1936). There is a large-scale debate between realist and idealist interpretations of the TM. Many textbooks today present the TM as a purely mathematical entity, an abstract mathematical idea.

Purely mathematical notions, such as sets of symbols, and functions from state–symbol pairs, replace Turing's scanner, tape, and punch-holes.[1] This conception has been criticized:

> Turing's bold innovation has been purified and rendered into the conventional coin of mathematics. Turing machines are no longer objects located in time and space, and subject to cause and effect. The paper tape, and the punched patterns that cause the machine to act in certain ways, are gone.
>
> (Copeland 2017, 54)

Turing-machine realists regard the mathematics

> merely as a useful formal representation of a Turing machine. But, just as a mathematical representation of digestion should not be confused with the process of digestion itself, so too the mathematical representation of a Turing machine must not be confused with the thing that is represented—namely, an idealized physical machine.
>
> (Copeland 2017, 54)

According to Copeland and Shagrir (2011), Turing himself was not a subscriber of the idealist, or purist, version of the TM.

Supporting a realist or purist conception of TM has important theoretical consequences in the conception of computation. As Copeland and Shagrir (2011) point out, the purist version of the TM features an internal version of the computation. A function f is computable by a machine in the internal sense just in case the machine is able to produce f (n) for any argument n in the domain, indicating that the value f (n) has been produced (printed in the output square) either by halting once the value is printed or by some other means.

> When computing in the internal sense, the only restriction on methods that may be used to indicate that the value has been produced is this: the method must involve no appeal to the behaviour of some device or system that is external to the specification of the machine—such as a clock.
>
> (Copeland and Shagrir 2011, 229)

A realist conception of the TM instead supports an external version of computation. A function f is computable by a machine in the external sense just in case the machine is able to produce f (n) (for any argument n in the domain) by performing, or failing to perform, some prespecified action during a pre-specified time-interval (open or closed) that is delimited by reference to the activity of some entity external to the machine.

My thesis is that a semiotic interpretation of the TM gives us the tools to overcome the dualism between purism and realism and internal and external computation. I suggest that there is an iconic relationship between the two, in the sense that the semiosis mediates between the machine and the abstract structure and allows the application of one to the other. Semiotic relationships mediate between matter and idea, technology and abstract structures.

The TM uses symbols and connections between symbols (1, 0) and manipulates them following several rules. Turing proposes a dyadic interpretation of the semiotic process that occurs in the machine. Everything is reduced to the dualism of input/output, data in and data out. The concepts of algorithm and data are structured according to this dualism. There are no data in an abstract sense; data are always connected to operations, that is, algorithms, or processes in general, because not all processes in a computer are algorithms. Data are the assumption (input) or the result (output) of a process.

What happens if we apply another conception of sign to the TM? How does this application transform our concept of the TM? According to Peirce, semiosis is not a dyadic process, nor can it be reduced to a dyadic process.

Peirce claims that

> It is important to understand what I mean by semiosis. All dynamical action, or action of brute force, physical or psychical, either takes place between two subjects [whether they react equally upon each other, or one is agent and the other patient, entirely or partially] or at any rate is a resultant of such actions between pairs. But by "semiosis" I mean, on the contrary, an action, or influence, which is, or involves, a cooperation of three subjects, such as a sign, its object, and its interpretant, this tri-relative influence not being in any way resolvable into actions between pairs.
>
> (Peirce 1931–1958, henceforth quoted as CP 5.484)

According to Peirce, semiosis is a type of relationship that includes three subjects: sign, object, and interpretant. This definition must be carefully analyzed. I emphasize two aspects: (a) the non-reducibility of the triad, which is the so-called "reduction thesis," and (b) the concept of infinite semiosis. Both cannot be understood without referring to the general structure of Peirce's thought—the systematic character and plurality of dimensions (logical-mathematical, pragmatist, phenomenological, and metaphysical-idealist) are strictly interconnected.

The first aspect is very complex and has sparked enormous debate in the literature. Peirce's thinking is deeply triadic; this structure has been extensively analyzed by Spinks (1991). The reduction thesis, according to which true triadic relations cannot be analyzed in monadic or dyadic terms, has been discussed in Burch (1993), Anellis (1993), and Hereth Correia et al. (2006). I do not want to fully analyze this debate. I limit myself to underscoring that this thesis has its roots in the phenomenological and metaphysical distinction between primality (possibility), secondness (existence), and thirdness (reality, regularity, habits). In Peirce, there is a fundamental equivalence: thirdness = reality = habits = mind = semiosis. Mind is an iterative tendency within matter. This iterative tendency is semiosis, which therefore is not a simple human mental operation, but a structure of reality that involves humans and nonhumans. For Peirce, the mind, as a set of habits, is thirdness and thirdness is semiosis. Consequently, "the mind is of the same nature throughout the universe—albeit differently diffused in the various parts of it. The mind of man is not essentially different from the mind of nature—because otherwise he could not know it" (Fadda 2013, 166; my translation). It is an anti-anthropocentric and anti-psychological conception of the mind: "Although Peirce does not trust psychological explanation, he realizes the importance of human activity in signing, but he also recognizes that there is a relationship which functions for, not just in, the human mind" (Spinks 1991, 53–54).

Thanks to this ontological perspective, Peirce overcomes Saussure's dualism, which distinguished two opposing but connected dimensions in the sign: the ideal meaning (signified) and the material signifier, which are united by convention. As I mentioned earlier, Peirce offers us a much more complex triadic model. Let us read the classic definition:

> A sign, or *representamen*, is something which stands to somebody for something in some respect or capacity. It addresses somebody, that is, creates in the mind of that person an equivalent sign, or perhaps a more developed sign. That sign which it creates I call the interpretant of the first sign. The sign stands for something, its object. It stands for that object, not in all respects, but in reference to a sort of idea, which I have sometimes called the ground of the representamen. "Idea" is here to be understood in a sort of Platonic sense, very familiar in everyday talk; I mean in that sense in which we say that one man catches another man's idea, in which we say that when a man recalls what he was thinking of at some previous time, he recalls the same idea, and in which when a man continues to think anything, say for a tenth of a second, in so far as the thought continues to agree with itself during that time, that is to have a like content, it is the same idea, and is not at each instant of the interval a new idea.
>
> (CP 2.228)[2]

The representamen is the first element of the sign relation. The term *representamen* derives from the Latin *re-praesento* (it represents, recalls, puts under the eyes; the sign *stands for something*). *Representamen* is therefore everything that makes something else present in some way. Peirce claims:

> To stand for, that is, to be in such a relation to another that for certain purposes it is treated by some mind as if it were that other. Thus a spokesman, deputy, attorney, agent, vicar, diagram, symptom, counter, description, concept, premises, testimony, all represent something else, in their several ways, to minds who consider them in that way. When it is desired to distinguish between that which represents and the act or relation of representing, the former may be termed the "representamen," the latter the "representation."
>
> (CP 2.273)

In the Peircean view, the sign itself implies three aspects: (a) *otherness*, because it represents something other than itself; (b) *absence*, because this something else is absent; and (c) *futureness*, because the sign opens to a further temporal dimension, to a perspective that goes beyond the immediate present (Kohn 2013, 34). Otherness, absence, and futureness: this is the profound structure of the sign. However, these aspects of the sign could not be fully understood without referring to the other two components of the semiotic relationship: the interpretant and the object.

Many books schematize the sign–interpretant–object relationship with a triangle inspired by Ogden and Richards' famous semiotic triangle (Ogden and Richards 1923), which includes concept–sign–referent. However, this operation may perhaps only be useful in the very first place, but it is mostly misleading, because such assimilations make one lose the peculiar characteristics of the Peircean model (see Fadda 2013, 170). The Ogden–Richards triangular model "is not triadic at all but is reduced to dyadic combinations (the sign-concept and the concept-referent relationship, while the one between sign and referent is only indirect)" (Fadda 2013, 170; my translation). Furthermore, Peirce has never explicitly spoken of a triangle, except in a few passages in which he proposes the image of a triangle with the tip pointing downward (Fadda 2013, 170).

There is another reason why the triangular model cannot be accepted, however. The reason is that the triangular model misses an essential trait of Peircean semiosis, that is, its dynamism. Talking about the object, Peirce distinguishes between the immediate object and the dynamic object.

> Signs can be classified on the basis of the characters which (1) they, (2) their immediate and (3) their dynamical objects, and their (4) immediate, (5) dynamical and (6) final interpretants possess, as well as on the basis of the nature of relations which (7) the dynamical objects and the (8) dynamical and (9) final interpretants have to the sign and which the (10) final interpretant has to the object. These ten divisions provide thirty designations for signs (each division being trichotomized by the categories, First, Second and Third). When

> properly arranged, they are easily shown to yield but sixty-six classes of possible signs.
>
> (CP 4.536)

The concept of immediate object is inseparable from that of interpretant. The immediate object is the object as represented by the chain of signs. The interpretant is nothing more than a new sign created by the first one. The main function of every sign is to create another sign that interprets it, in the sense that it makes explicit its relationship with the object. This relationship is the immediate object—how a sign stands for the object. Consequently, for Peirce, there cannot be an isolated sign. Each sign is a variable that must be determined by another sign, and so on. "Admitting that connected Signs must have a Quasi-mind, it may further be declared that there can be no isolated sign" (CP 4.551). In other terms, each sign must produce another sign to be able to represent the object. Interpreting means creating a new sign. The semiotic process is always unstable and evolving. There is never a definitive, ultimate sign where the process stops completely. The dynamic object is instead the source and the goal of the process, its terminus ad quo and ad quem. The entire semiotic process is a movement of infinite approximation to the dynamic object; immediate objects are only approximations of the dynamic object. Nevertheless, for Peirce, the dynamic object is not the Kantian noumenon; the dynamic object is perfectly knowable, even if only through a process of continuous approximation through the sign.

This is the infinite semiosis, as Peirce explains:

> A sign stands for something to the idea which it produces or modifies. Or it is a vehicle conveying into the mind something from without. That for which it stands is called its object; that which it conveys, its meaning; and the idea to which it gives rise, its interpretant. The object of representation can be nothing but a representation of which the first representation is the interpretant. But an endless series of representations, each representing the one behind it, may be conceived to have an absolute object at its limit. The meaning of a representation can be nothing but a representation. In fact, it is nothing but the representation itself conceived as stripped of irrelevant clothing. But this clothing never can be completely stripped off; it is only changed for something more diaphanous. So there is an infinite regression here. Finally, the interpretant is nothing but another representation to which the torch of truth is handed along; and as representation, it has its interpretant again. Lo, another infinite series.
>
> (CP 1.339)

> Sign is anything which is related to a Second thing, its Object, in respect to a Quality, in such a way as to bring a Third thing, its

> Interpretant, into relation to the same Object, and that in such a way as to bring a Fourth into relation to that Object in the same form, ad infinitum. If the series is broken off, the Sign, in so far, falls short of the perfect significant character. It is not necessary that the Interpretant should actually exist. A being in futuro will suffice.
>
> (CP 2.92)

A caveat is necessary. The interpretant is neither the meaning of the sign nor an interpreter—on the contrary, the interpreter is a type of interpretant. To understand this point, it must be emphasized that in Peirce, the notion of meaning is doubled: on the one hand, the meaning of a sign is the dynamic object, so each sign tells one something more about the object in a process of continuous approximation; on the other hand, in a pragmatist sense, the meaning of a sign is the way in which it transforms habits. Therefore, the interpretant is the meaning of a term to the extent that it allows the relationship between that term and the dynamic object. Peirce's perspective is realistic, in the sense that the dynamism of semiosis is not chaos but is regulated by regularities, habits. The dynamism of semiosis is at the same time "attracted" by the dynamic object and regulated by habits—for example, the dispositions of the one (human and nonhuman) who receives the sign.[3]

The interpretant is a sign produced by another sign in relation to the dynamic object and according to a certain number of dispositions, or habits—regularity, laws. For this reason, every semiotic process implies an infinite regression. The paradoxical character of Peirce's approach is only apparent, as Eco (1975) explains:

> However paradoxical the solution may seem, infinite semiosis is the only guarantee of a semiotic system capable of explaining itself in its own terms. The sum of the various languages would be a self-explanatory system, or a system that is explained by successive systems of conventions that clarify each other.
>
> (122; my translation)

Let us go back to Turing now. The TM is a linear model where the meaning of the signs (1, 0) is given by the rules without any further mediation (Figure 5.1).

Reinterpreting this model in Peircean terms, we have this situation (Figure 5.2):

In the Semiotic Turing Machine (STM), the meaning of the signs (1, 0) is established not only by the rules, but also by the set of interpretants (int^n). For Peirce, the interpretant is not necessarily a rule, or a set of rules, nor necessarily human or conscious. It is a sign that

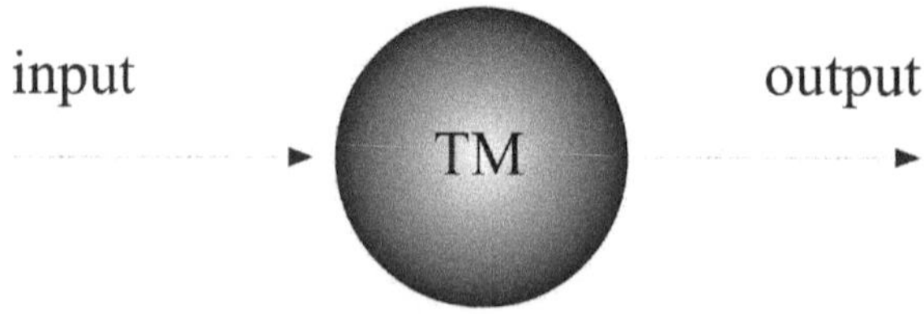

Figure 5.1 The Turing information model. It is a constructionist model. Information is the answer to the question: "How is X built?" "What is the procedure that allows to construct (and therefore to know) X?" To any X the TM allows to connect a P (x) that is the procedure that allows to construct X. This is computation.

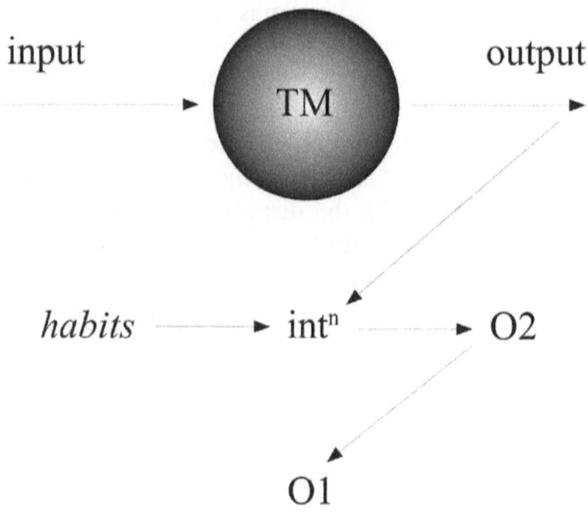

Figure 5.2 The reinterpretation of the Turing model through Peirce.

varies continuously, forming a series of approximations (the immediate object, O2) to the dynamic object (O1), which in this case is the number or function. Furthermore, the interpretant refers to one or more habits. The variation of the interpretant is conditioned by the habit; even in this case, we are not forced to identify habits and rules, in the sense that rules are forms of habit, but the reverse is not necessarily true. The habit is any generalizing trend:

> If we now revert to the psychological assumption originally made, we shall see that it is already largely eliminated by the consideration that habit is by no means exclusively a mental fact. Empirically, we

find that some plants take habits. The stream of water that wears a bed for itself is forming a habit.

(CP 5.492; see also CP 2.643, 2.148, 5.480)

The habit is a normative and ontological category.

We must then add to the three previously distinguished dimensions of the sign (otherness, absence, and futureness) three further dimensions: (a) the possibility, that is, the possibility that that sign can be interpreted in that chain of signs; (b) the existence, that is, the physical presence of the sign; and (c) the regularities involved by the semiotic process in the double sense (c1) of the effect of the sign in the long term, or how that sign influences the development of the chain of signs and (c2) of the human and nonhuman regularities that determine the associations of that sign, that is, how that sign is connected to the other signs, its interpretants, which constitute the spectrum of its possible interpretations.

TM presupposes STM. This latter has two crucial differences compared to the first: (a) the operation is always the creation of an interpretant that is governed by local habits, rules, and the dynamic object; (b) the result is not a sign, an output, but the production of a new habit, or lasting transformation of an already existing habit—what Peirce calls the "final logical interpretant." In Latourian terms, the result is the transformation of power relations, that is, to what extent we are able to decompose that number or function in order to prove its computability; how far that number or function resists the work of TM (see *Irreductions* 1.1.5, 1.1.6, and 1.1.15). If, as Latour says, each actant defines its inside and outside in constant negotiation with other actants, *then* TM can be considered as an actant who fights precisely to make this distinction. In other words, still following Latour's terminology, *computation is translation*, that is, relationship between forces and the construction of an identity (see *Irreduction* 1.2.1). The forces facing each other on this battlefield are diagrams and numbers. To these are subsequently added other actants, such as algorithms, programming languages, and hardware, as we will see in the next section. In other words, TM is the demonstration/discovery that even numbers and mathematics are translation, conflict, and negotiation. However, I do not follow Latour when he thinks that there is nothing among the forces: "There is nothing between incommensurable and irreducible forces: no ether, no instantaneousness" (Latour 1988, 162). How could forces communicate if there were not an elementary, material semiosis to constitute the environment in which they move?

What does STM add to TM? TM alone cannot solve the halting problem. The halting problem represents the barrier against which TM crashes. STM connects the purely formal language of TM to a network of icons and indexes that define TM's umwelt. This means that if the halting problem cannot be solved on the symbolic level, it can be solved on other levels, such as the pragmatic or semantic one. Turing proved that a general algorithm to solve the halting problem cannot exist. However, we can solve the problem, i.e., establishing whether the program will finish running, or continue to run forever, in a pragmatic way, i.e., analyzing the umwelt of the program, its resources and the environmental constraints.

A Peircean Theory of Computation

In this section, I apply Peirce's triad of icon, index, and symbol to the TM. I intend to show that the TM is based on a complex web of iconic relations. I also introduce the relationship between computation and algorithm and show that it is a semiotic relationship. This is an essential point for proving the central thesis of this paper: an explanation of computation that does not consider the semiotic aspect of computation is incomplete. Without semiotic abilities, the TM cannot work.

There is a question that arises from the previous considerations: what characterizes the TM as a semiotic process? The only result we have achieved so far has been to criticize the conception of the sign implicit in TM and to propose the model of Peirce's infinite semiosis. Now we need to understand how semiosis develops in TM.

I want to start from a very simple non-technical definition: computation means decomposition (see Primiero 2020, Chapter 4). Computing X means decomposing X into a series of symbolic operations until X is reduced to very simple and self-evident computations. The relationships between the operations can be described in different ways starting with the basic operations. This is what the theory of recursive functions states; a recursive function (primitive or partial) is reducible to/derivable from a small group of very simple operations such as the zero function, identity functions, successor, constant, composition, primitive recursion, and so on. The TM does exactly this: X is computable when it is reducible to a set of elementary operations that stand among themselves in specific relations. If this decomposition/construction is possible, X is computable; otherwise, it is not. The TM and recursion theory do the same thing, even differently—decompose X into a procedure. The Church–Turing thesis claims that recursive functions are identical to the set of functions $f\colon N \rightarrow N$ that can be mechanically computed, that is, are programmable on some computer. The set of partial recursive functions is identical to the set of Turing-computable functions.

Let us assume this basic definition: to compute means to reduce X to a controllable procedure. The controllability derives mainly from isomorphism, that is, from the fact that all computable objects have the same structure, that is, those basic and self-evident operations and their relations. Therefore, we can claim that the TM and recursion theory are essential tools for studying algorithms and their complexity.

Now, Peirce distinguishes three types of signs: icons, indices, and symbols. In the icon, there is a relationship of similarity between sign and object; iconic signs are divided into three classes: images, diagrams, and metaphors. The index, on the other hand, expresses a relationship of space–time contiguity between sign and object; the index has a current connection with its object. In the symbol, on the other hand, the relationship between sign and object is of a conventional nature; the sign implies a social habit that regulates its relationship with the object. (This is obviously a very general description; for more detail, see Spinks 1991, Chapter 3.)

It is important to emphasize that Peirce has always been very critical of the category of similarity and of the attempt to define the icon only in terms of similarity.[4] It is not the similarity that defines the icon, but its epistemological function. As Stjernfelt (2000) shows, the crucial feature of the icon in Pierce is that of being the only sign capable, starting from its simple observation, of discovering something new about the object represented. The icon has an essential heuristic function. An icon that is simple similarity, but reveals nothing of the object, is not a real icon. According to Peirce, any form of reasoning is a manipulation of icons, that is, the observation and transformation of icons. By experimenting on icons, that is, modifying and transforming them, we can discover new information on the represented object, new relationships in particular—Plato's experiment in *Meno* is a classic example.

The concept of icon is essential in mathematics. For Peirce, in fact, mathematics has two characteristics: (a) it is the science of what is deductively possible, the fundamental basis of any knowledge—the mathematician develops deductions starting from certain hypotheses; (b) it is an iconic science, in the sense that it develops its own deductions starting from signs that are specific types of icons, that is, diagrams: drawings (geometry) or alphanumeric notations (algebra). The diagram represents its object through a skeleton-like sketch of relations.

> A *diagram* is an icon or schematic image embodying the meaning of a general predicate; and from the observation of this icon we are supposed to construct a new general predicate.
>
> (Peirce 1976, 238)

> To begin with, then, a Diagram is an Icon of a set of *rationally* related objects. By rationally related, I mean that there is between them, not merely one of those relations which we know by experience, but know not how to comprehend, but one of those relations which anybody who reasons at all must have an inward acquaintance with.
>
> (Peirce 1976, 316)

Thus, the inclusion of algebra, syntax, and the like in the icon category is due to their diagrammatic properties. In other words, according to Stjernfelt (2000, 365), "the diagram is so to speak the redrawing of an icon in terms of a priori relations among its parts." An original diagrammatization precedes any other type of icon: images and metaphors. "The diagrammatic way of interpreting an icon seems central as soon as any part of the internal mereological structure of the icon is taken into consideration" (Stjernfelt 2000, 361).

The mathematician works on diagrams. These diagrams do not refer to an individual object, but to classes of possible objects (numbers or geometric figures) identifying and representing their common qualities. The mathematician has two possibilities: either it limits itself to observing the diagram and draws all the conclusions it can draw from it, or it modifies the diagram according to the rules it has available and thus obtains a new diagram that can reveal some new truth that cannot be drawn from the previous diagram. In this second case, says Peirce, we have a theorem:

> A *theorem*, as I shall use the word, is an inference obtained by constructing a diagram according to a general precept, and after modifying it as ingenuity may dictate, observing in it certain relations, and showing that they must subsist in every case, retranslating the proposition into general terms. . . . A theorem regularly begins with, 1st, the general enunciation. There follows, 2nd, a precept for a diagram, in which letters are employed. Then comes, 3rd, the ecthesis, which states what it will be sufficient to show must, in every case, be true concerning the diagram. The 4th article is the subsidiary construction, by which the diagram is modified in some manner already shown to be possible. The 5th article is the demonstration, which traces out the reasons why a certain relation must always subsist between the parts of the diagram. Finally, and 6thly, it is pointed out, by some such expression as . . . the usual Q.E.D., or otherwise, that it was all that it was required to show.
>
> (Peirce 1976, 238)

Based on Peirce's theory of diagram, I claim that the condition of the computational decomposition is the existence of a double iconic relationship: (a) that between the diagrams used by the mathematician and the dynamic object (numbers or functions) and (b) that between the different steps of the decomposition of the diagram according to certain rules. The first point is confirmed by the diagrammatic nature of mathematical knowledge. The second is characteristic of computation: there is a diagrammatic relationship between all the steps of the computational decomposition because they all have the same structure, that is, they can all be reduced to the same fundamental structure—the basic operations: successor, constant, composition, primitive recursion, and so on and especially the most basic functions: the zero function and the identity function (Boolos et al. 2007, 63–65). Computation is a form of manipulating diagrams to discover something new about the dynamic object.

What kind of relationship is represented in b? The isomorphism, that is, a transformation that preserves the mapping between two structures. The kind of isomorphism involved in computation is an automorphism, that is, an isomorphism between a structure and itself. The diagram is "broken down" in a symmetrical, isomorphic way. The theory of recursive functions claims that we can define recursive functions starting from a very small number of elementary, intuitive operations. All can be reduced to these basic structures; their intuitive computability must be preserved during the construction of new functions. The TM presupposes and reproduces this kind of diagrammatic-automorphic chain of signs, which is based on some habits. These habits are (a) the regularities present in a certain group of objects, the numbers, (b) the rules of the TM, and (c) the inferential abilities of the human subject.

What is the relationship between this automorphism inside the chain of signs and the dynamic object? From a Peircean perspective, I would say that there is a relationship of reciprocal construction; the automorphing sign chain defines—and is defined by—the dynamic object (numbers) at the same time. The process of continuous approximation to the dynamic object can be confirmed or denied, and therefore proceed in another direction or stop.

Thus, from the point of view of Peirce's semiotics, the TM is a set of diagrammatic transformations defined by three groups of regularities, or habits. The diagrammatic automorphism guarantees the controllability of the chain of signs and the discovery of new computable functions. Claiming the importance of icons and diagrams for mathematical thinking, in line with Peirce, means overcoming a rigidly formalist and axiomatic perspective of mathematical work and instead showing its intuitive and creative background. It also means showing the limits of formalization in mathematics (Cellucci 2019, 593).

We must now introduce an essential distinction between computation and algorithm. This also poses another important problem: What happens when the dynamic object changes? What happens when the semiotic chain of the computational process is connected to another semiotic chain?

TM is the normative point of reference for the computational process; it is a method of analysis and control used to evaluate the complexity of concrete, individual computations. Turing himself described his 1936 paper as "an investigation of the theoretical possibilities and limitations of digital computing machines" (cited in Copeland 2017, 56). The main goal of computer programming is to translate algorithms into computational terms; the TM must be seen as the model we use to evaluate the complexity of our algorithms. The RAM model, which is the fundamental model used in computer science to design algorithms, is in fact based on the TM.

The notion of algorithm is broader than the concept of computation and the TM. In general, an algorithm proposes a method to solve a given problem.

> An algorithm is a procedure to accomplish a specific task. An algorithm is the idea behind any reasonable computer program. To be interesting, an algorithm must solve a general, well-specified problem. An algorithmic problem is specified by describing the complete set of instances it must work on and of its output after running on one of these instances. This distinction, between a problem and an instance of a problem, is fundamental.
>
> (Skiena 2008, 3)

We can interpret the relationship between algorithm and the TM in terms of Peirce's infinite semiosis (see Figure 5.3). Following Latour (1994) and Akrich and Latour (1992), an algorithm can be considered a set of practices and provisions, a "program of action." This program (PA1) is the dynamic object to which a series of interpretants refer

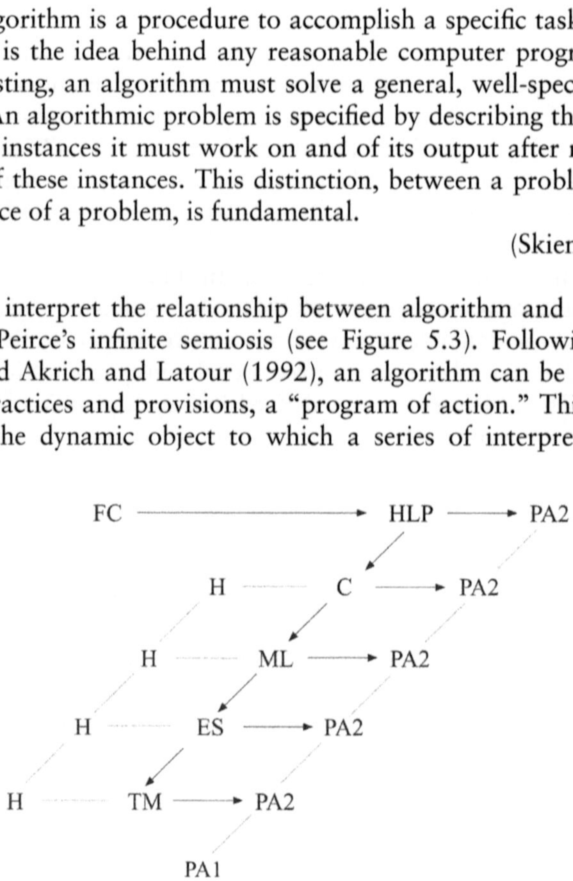

Figure 5.3 The semiotic structure of TM. Different systems of interpretants mediate between the TM and the algorithm: flow diagrams (FC), high-level programming languages (HLP), and compilers (C), that is, low-level programming languages that in turn mediate between HLP, the machine language (ML), and the electrical signals (ES). The program (PA1) is the dynamic object to which a series of interpretants refer through the immediate object (PA2). The process is entirely regulated by habits (H).

through the immediate object (PA2). The TM is an interpretant of the algorithm because it tells us something more about the PA, that is, its computational complexity. Different systems of interpretants mediate between the TM and the algorithm: flow diagrams (FC), high-level programming languages (HLP), and compilers (C), that is, low-level programming languages that in turn mediate between HLP, the machine language (ML), and the electrical signals (ES). HLP and C are mainly formed by symbols and indices, that is, by conventional human signs and by signs that express physical–temporal contiguities (for example, the connection between the instructions and the places in the memory where to take the data). C has a more indexical nature than HLP, which is more symbolic.

We thus arrive at a provisional conclusion of our semiotic reinterpretation of computation. Computation has a semiotic structure composed of a network of icons, indexes, and symbols that constantly translate into each other. We can distinguish at least four iconic levels: (1) the relationship between ML and ES, (2) the networks of diagrams in TM, (3) the relationship between TM and ML-ES, and finally, (4) the relationship between ML-ES and the PA. There is an iconic relation between ML and ES in the sense that ES must be a diagram of ML; the electrical signals must reproduce the same logical relations expressed by the code machine. The symbolic-level HLP and indexical-level C constantly mediate between the two iconic levels 1 and 2, transforming them mutually—they manipulate the icons to create something new. In fact, the goal of this mediation is to create two new iconic relationships, that is, 3 and 4. 3 is the iconic relationship between 1 and 2, in the sense that ML-ES must become a diagram of the TM—it should diagrammatically represent the TM, that is, present the same structure. 4 is the iconic relationship between 3 and PA, the dynamic object (Figure 5.4).

Computation takes place when all these semiotic relations happen. The algorithm and PA are then translated into computational terms.

Therefore, computation is not only a mathematical abstraction based on some rules or axioms but implies a very complex and dynamic network of changing semiotic relationships. Without semiotic relationships, we could not understand the mathematical process. Furthermore, without semiotic relationships, we could not even understand the relationship between the purist and the realist interpretation of the TM. Semiosis mediates between internal computation (TM) and external computation (PA). This semiotic analysis shows us that computation is a creative process based on the manipulation of icons, that is, diagrams. It is therefore a heuristic process that makes us discover something new about its object (PA) and the surrounding reality.

We have not considered a fundamental point, however. PA can change, in two senses. First, it can receive a different interpretation. This means that, in the development of infinite semiosis, PA can be

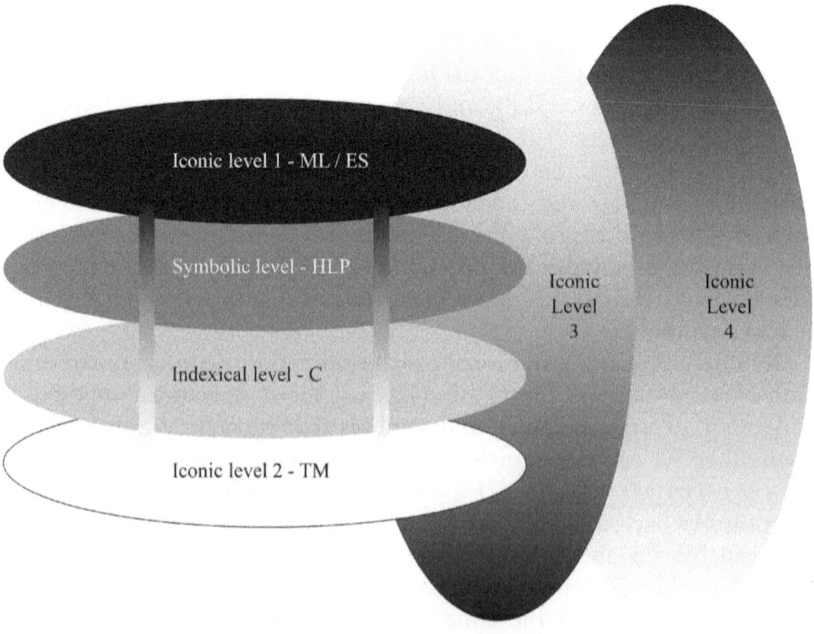

Figure 5.4 The semiotic structure of computation (STM). It is composed of several semiotic levels that interact with each other. The open and hybrid structure of the sign allows the STM to mediate between the ideal TM, which is a purely mathematical structure, and the real TM, which is a machine in a context.

connected to a different interpretant, which modifies its meaning. Semiosis undergoes an unexpected curvature. Second, quoting Akrich and Latour (1992), each PA is always in relation to an anti-PA, that is, an antiprogram. The relationship between PA and anti-PA is defined by the interpretation of the PA, that is, by the development of semiosis. This may seem strange if we stick to the situation, very simple and somewhat ideal, of the isolated programmer programming her/his laptop. Instead, we should think of the current AI systems, as capable of learning on their own from immense numbers of data and changing themselves. If what we said before is true, the PA changes constantly in contact with the real world.

Mechanistic Account of Computation and Semiosis

In this section, I want to analyze another conception of computation, namely, the mechanical explanation of computation as a physical system formulated by Piccinini (2015). As I will show, this explanation is incomplete if we do not introduce semiotic relations.

An objector might say that the conception of computation developed in the previous section is like what the literature calls semantic account. According to the semantic account, "a computation is a process that manipulates representations in an appropriate way" (Piccinini 2016, 205). According to the semantic account, "computations are individuated at least in part by their semantic properties" (205). According to a certain version of the semantic account (Shagrir 2006), if every physical state carries representation and information, then every physical system performs the computations constituted by the manipulation of its information-carrying states.

This objection is wrong: as said at the beginning, Peirce's semiotics does not presuppose a semantics. A Peircean semantics does not exist and this is for two reasons: (a) Peirce does not develop any systematic investigation of languages and (b) the Peircean notions that can make us think of a semantics, that is, interpretant and meaning, refer to two other concepts, which are semiosis and pragmatism (Fadda 2013, 185). This is by no means the same as excluding the notion of meaning from our explanation of computation; in fact, my thesis is that computation does not presuppose meanings, but produces them.

According to Piccinini (2015, Chapter 3), the semantic account is inadequate because it must necessarily use only general representations and these representations can be interpreted in very different ways. Consequently:

> For the same accidental representation may represent different things (including nothing at all) to different interpreters. As a consequence, a putative computation that is individuated by reference to the semantic properties of accidental representations may be taken by different interpreters to compute different things without changing anything in the process itself. Just as speakers of different languages can interpret the same string of letters in different ways, under the semantic account (plus the notion of accidental representation) different observers could look at the same activity of the same mechanism and interpret it as two different computations.
>
> (Piccinini 2016, 206)[5]

Another criticism that could be formulated against what we said in the previous section is that our conception of computation is too abstract and applicable to anything, and therefore useless. In other words, it is similar to the mapping account (Putnam 1967; Chalmers 2011). According to this conception of computation, anything that we can describe in computational terms, that is, as the TM, is a computational system.

This objection is incorrect. Connecting computation to Peirce's infinite semiosis does not mean identifying computation and semiosis, but to show how computation is related to infinite semiosis, or the semiosphere.

The mapping account is too vague: if everything is computation, nothing is computation: "To determine with some precision which class of systems can be described computationally to which degree of accuracy is difficult" (Piccinini 2016, 208). Furthermore, "the same physical system may be given many computational descriptions that are different in nontrivial respects, for instance because they employ different computational formalisms, different assumptions about the system, or different amounts of computational resources" (Piccinini 2016, 208). If everything can be described in terms of inputs and outputs, then the central question will be: Which input-output processes can be called computational in themselves?

The mechanistic account of computation comes from the rejection of the two previously mentioned approaches. According to the mechanistic account, computation is a type of mechanical system with specific characteristics. Piccinini (2015, Chapter 7) argues that a mechanical system is composed of three fundamental elements: working components, capacities assigned to the components, and organizational relations between the components. In other words, a mechanical system is a set of components and operations arranged in a certain way, that is, according to a certain level of organization. Piccinini distinguishes two types of mechanical systems: teleological and non-teleological systems. The first are organisms and artifacts. These systems, in fact, have purposes that are related to their functions. A living organism has the fundamental purpose of staying alive, for example. A car is meant to transport people from one place to another in a comfortable, safe, and fast way. These general purposes fall into a myriad of more specific purposes assigned to the different components of the system.

Piccinini distinguishes the computational system and the mechanical system. These are not the same thing; otherwise, any organism or artifact would automatically become a computational system, and this does not make any sense, as already mentioned. A computational system is a certain type of mechanical system. We cannot even identify the TM with the computational system because similar, non-digital systems are also computational systems. We need to give a broader definition of a computational system than TM. Piccinini claims that

> A computation in the generic sense, then, is a process defined by a general rule for manipulating some kind of vehicle based on differences between different portions of the vehicle along some dimension of variation. *A computing system in the generic sense is a mechanism whose (teleological) function is manipulating some type of vehicle* (digital, analog, or what have you) *in accordance with a rule that is general*—namely, it applies to all vehicles of the relevant kind—*and that depends on the input vehicles (and perhaps internal states) for its application.* A computational explanation in the generic sense,

> then, is a mechanistic explanation in which the inputs, outputs, and perhaps internal states of the system are medium independent vehicles, and the processing of the vehicles can be accurately captured by appropriate rules.
>
> (Piccinini 2016, 214)

Let us carefully analyze this definition. There are three aspects that characterize a computational system: (a) abstraction (vehicles and operations are medium-independent); (b) the general rules; and (c) the internal states of the system. The general rules organize the relationships between vehicles and functions in relation to the internal state of the system. A vehicle consists of spatiotemporal parts or portions; for example, a string of digits is a kind of vehicle that is made out of digits concatenated together.[6] The rules only concern the differences between vehicles and vehicle parts, not the qualities of the vehicles themselves. The key notion is that of a functional mechanism.

> A system X is a functional mechanism just in case it consists of a set of spatiotemporal components, the properties (causal powers) that contribute to the system's teleological functions, and their organization, such that X possesses its capacities because of how X's components and their properties are organized.
>
> (Piccinini 2015, 119)

According to Piccinini, this definition is not only the most appropriate to the way of thinking and working of engineers and computer scientists, but it is also applicable to the TM, in the sense that the TM is a type of functional mechanism. "A similar notion of functional mechanism applies to computing systems that are defined purely mathematically, such as (unimplemented) Turing machines" (Piccinini 2015, 119). The purpose of the computational system is to transform vehicles, or portions of vehicles, according to the general rules.

Now, we could describe a computational system in terms of the ANT and Latour's semiotic approach. However, this operation would greatly limit our investigation because it would imply the reduction of the computational system to a single narrative model, that of Greimas. Instead, it is possible to maintain the same methodological approach of Latour based on the principle of human–nonhuman symmetry (Latour 1987, 1988) but using another type of semiotics inspired by Peirce. Peircean semiotics seems to me a more adequate model because (a) it is more flexible and does not reduce every collective or network to a single logical or narrative system and (b) it has empirical roots, as the developments in biosemiotics demonstrate (Sebeok 2001c). Peirce's semiotics and its development in biosemiotics represent a radical overcoming of any possible dualism between human and nonhuman.

The mechanistic account of computation involves several semiotic relationships. The relationship between operations and goals is indeed semiotic. The operation is a sign of the goal, in the sense that the first refers to the second and asks for an interpretant. The goal is the (immediate and dynamic) object of the chain of signs.

Furthermore, Piccinini misses a fundamental aspect: (a) function, (b) objective, and (c) their relationship does not lie in the object. If I open a laptop, I do not find functions and objectives, I find only hardware and electrical impulses. The function of my laptop's fan is not in the fan as a material object; it is generated by semiosis because the fan takes on a function and a goal only within a network of relationships, postponements, and absences—the fan acts when the right temperature is missing. The fan of my laptop would not exist if there were not this absence (the absence of the right temperature) that constitutes it. "This something-not-there permeates and organizes what is physically present in these phenomena" (Deacon 2011, 9). Absence is not materially present but acts causally—it turns on the fan. The movement of the fan is a sign of the absence of the right temperature—it is a semiotic relationship in Peirce's terms—and the effect of this absence at the same time. This is what Deacon calls "without-ness," an absence that is constitutive, acts, and defines a physical phenomenon—a real heresy for mainstream science. "What is absent matters" (Deacon 2011, 3). This is "a defining property of life and mind" (Deacon 2011, 3). Absences such as function, goal, or meaning should not be thought of as metaphysical entities; they act materially because they have effects, and they manage to do so because they are generated by material, physical processes. Therefore, paradoxically, a mechanistic explanation of computation must include an absential account—it can be a simple mechanistic account.

This is also a very important point with respect to our critique of the Latourian semiotic method. Latour's ANT fails to explain the absences in the collectives; the associations between actants pass only through presences confirmed by "tests of strength" (Latour 1988). Latour and ANT miss Deacon's revolution, that of introducing into our way of seeing the world "a form of causality dependent on specifically absent features and unrealized potentials" (Deacon 2011, 16).

The notion of goal itself is very complex. The goal is not included in the operation and cannot be defined starting from the operation because the same operation can be interpreted in very different ways, that is, have very different purposes in the same system. For example, the car's goal of transporting humans from one destination to another quickly, comfortably, and safely involves a number of sub-goals (e.g., allowing adequate seat adjustment, air conditioning, and parking sensors) that can be implemented by many different operations at the same time—Latour would say: programs, sub-programs, and counter-programs. Coordinating these operations in relation to the purposes requires an interpretative capacity, that is, a selective ability: to choose the best operations to achieve certain purposes—in semiotic terms: to produce an interpretant for a sign, this is the interpretation in the Peircean sense. Furthermore, the same operation (for example, the activation of the parking sensors) can have different goals in the same system depending on the situation (to facilitate parking or avoid collision with a vehicle that is too close). The relationship between operations and goals is not 1–1, but n–n. It is not linear but reticular.

An objection could be that the goals and their architecture are already included in the general rules that define the organization of the system and that therefore the relationship between operations and goals is defined by the general rules; there is no need to add anything else. However, the objection does not consider a decisive aspect: the goals cannot be included in the rules because they can change. In every artifact there is a stratification of goals: there are more elementary, almost "natural" goals, which are included in the rules, such as "to manipulate the components according to the general rules," but there are also more complex goals, such as organizing pay slips in a company with thousands of employees. Elementary goals come from the design of the artifact. Complex purposes, on the other hand, come from the evolution of the system and from the relationship with the surrounding environment. Complex goals impact the most basic purposes and can change the design. This happens in the most complex artifacts, especially in computational systems. The relationships between the goals can be described in semiotic terms; a goal can be an icon or index of a series of sub-goals or imply a symbolic relationship with a series of even more complex goals. A computational system that cannot recognize these relationships does not work.

Another important semiotic relationship in a computation system is that between operation and the internal state of the system is semiotic, in the sense that the operation is a sign of a particular internal state of the system, that is, of a specific relationship between the components. In this case, the semiotic relationship can be interpreted in two ways: on the one hand, the operation is the expression or result of a certain state of the system; on the other hand, the operation induces or causes a certain state of the system, in the sense that it causes a modification of the previous state. For the TM, for example, it is essential to recognize the relationship between an operation and the change of state of the system; this recognition implies an interpretative capacity, a symbolic competence—in Peirce's terms, the ability to produce an interpretant of the dynamic object.

There are two other semiotic relationships that properly characterize a computational system as such and both have to do with a key notion for semiosis, namely, the difference:

- The relationship between the spatial arrangement of vehicles and function. The differences between vehicles "make a difference" quoting Bateson (1972, 459), in the sense that they have a semiotic effect—they are signs of functions.

- The relationship between system and media. Here we have the inverse process: the elimination of differences. As Piccinini explains, a computational system is medium-independent. Medium independence implies multiple realizability. "A property is multiply realizable just in case it can be fulfilled by sufficiently different kinds of causal mechanisms" (Piccinini 2015, 122). The computation system eliminates the differences between the various mediums, putting them all on the same level. The absence of differences is a sign of multiple possible achievements—it creates new possibilities.

Absence and presence of difference are a sort of "zero degree" of semiosis in computational systems. Here we also find a determining aspect of Peirce's semiotic conception: the dynamism of semiosis. A computational system must possess a basic semiotic capacity, that is, an ability to interpret the differences.

Design, engineering, and programming do not explain everything that goes on in the system. The behavior of the system cannot be fully explained/reduced from its project/program. There are semiotic relationships between the parts of the system which cannot be programmed, but which are presupposed by the programming itself. The error, the breakdown, occurs when these semiotic relations are broken. Using ANT terminology (Akrich and Latour 1992), the inscription, that is, in the passage from the design to the setting, is not an exact translation; the setting modifies the design by adding to it something that the design did not foresee and that cannot be explained by the design—nor by the redesign cycles. Each inscription is a crisis, a test of strength. For this reason—following Latour—the description, that is, the extraction of the script, is necessary. The elaboration of the script must therefore be based on the analysis of the semiotic relationships implicit in the setting. As Latour claims, we must have the humility to follow the process of translation from design to setting, analyzing the intertwining of mediators and tests of strength (*Irreductions*, 2.6.1).

Conclusions

The thesis of the paper is that semiotic processes are intrinsic to computation and computational systems. An explanation of computation that does not take this semiotic dimension into account is incomplete. I have tried to analyze these semiotic relationships as rigorously as possible.

The result of this research is twofold:

(a) If a computational system implies specific semiotic relations, this means that it is part of the larger phenomenon of semiosphere, and therefore is in profound continuity with other forms of semiosis. These forms of semiosis can be not only biological, the so-called biosemiosis, but also cultural semiosis. Therefore, if, as Kohn (2013) states, semiosis is something that goes beyond the boundaries of the human and involves all living beings from their genetic structure (Barbieri 2015) to their behavior (Marrone and Mangano 2018), then computational systems play an autonomous and creative role in the construction and conservation of the semiosphere.

(b) The semiotic interpretation of computation allows us to overcome the dualism between the purist version and the realist version of computation. Semiosis mediates between the inside of the TM and its surrounding environment.

Understanding the implicit semiosis in computational systems is a crucial condition for understanding and interpreting digital technology and AI. This means also overcoming a rigidly reductionist view of these machines as inert objects and showing how they can produce interpretants and meanings that can transform semiosis among living beings.

Notes

1 Here a purist description of TM: "A Turing machine is a specific kind of idealized machine for carrying out computations, especially computations on positive integers represented in monadic notation. We suppose that the computation takes place on a tape, marked into squares, which is unending in both directions—either because it is actually infinite or because there is someone stationed at each end to add extra blank squares as needed. Each square either is blank, or has a stroke printed on it. (We represent the blank by S0 or 0 or most often B, and the stroke by S1 or | or most often 1, depending on the context.) And with at most a finite number of exceptions, all squares are blank, both initially and at each subsequent stage of the computation. At each stage of the computation, the computer (that is, the human or mechanical agent doing the computation) is scanning some one square of the tape" (Boolos et al. 2007, 25).

2 For a comment on the passage, see Spinks (1991, 52–53).

3 I do not want to analyze the oscillation, very present in Peirce's texts, between a psychological reading and a communicative and non-psychological reading of the concept of semiosis. See Fadda (2013, 173–175).

4 I cannot analyze here, for reasons of space, on the debate on iconism in 1970–1980 semiotics. I only mention the famous criticisms of Goodman and Eco. I quote, in response to these criticisms, the position of Deacon (a long but important passage): "When we apply these terms to particular things, for instance, calling a particular sculpture an icon, a speedometer an indicator, or a coat of arms a symbol, we are engaging in a sort of tacit shorthand. What we usually mean is that they were designed to be interpreted that way, or are highly likely to be interpreted that way. So, for example, a striking resemblance does not make one thing an icon of another. Only when considering the features of one brings the other to mind because of this resemblance is the relationship iconic. Similarity does not cause iconicity, nor is iconicity the physical relationship of similarity. It is a kind of inferential process that is based on recognizing a similarity. As critics of the concept of iconicity have often pointed out, almost anything could be considered an icon of anything

else, depending on the vagueness of the similarity considered. The same point can be made for each of the other two modes of referential relationship: neither physical connection nor involvement in some conventional activity dictates that something is indexical or symbolic, respectively. Only when these are the basis by which one thing invokes another are we justified in calling their relationship indexical or symbolic" (Deacon 1997, 71).

5 Piccinini does not deny that computational systems can be semantic. He only states that these systems do not presuppose semantics. "Of course, many (though not all) computational vehicles do have semantic properties, and such semantic properties can be used to individuate computing systems and the functions they compute. The functions computed by physical systems that operate over representations can be individuated either semantically or non-semantically; the mechanistic account provides non-semantic individuation conditions" (Piccinini 2015, 118).

6 The vehicles are not exactly identical to the components; vehicles are not objects with a fixed state, but concrete or abstract variables; see Piccinini (2015, 119).

6 AI, Psychoanalysis, and the Critique of Identity

What is identity in psychoanalysis? Why is the critique of identity in psychoanalysis important for understanding AI? These are the two questions at the heart of this chapter. The main thesis is that AI has to do with human identity and the identification processes that constitute it. The first part of the chapter will be dedicated to the first question. I will investigate the main points of the critique of identity in psychoanalysis and how psychoanalysis can provide important conceptual resources for the field of identity studies. Then, in the following parts of the chapter, I will answer the second question.

I will show how, from a psychoanalytic and sociological point of view, AI can be considered both the cause and the effect of a crisis of identity in contemporary post-Fordist capitalism. To describe this crisis of contemporary identity, I will use three concepts taken from sociology: social acceleration, new individualism, and capitalist realism. My thesis is that these three concepts define a general crisis of the symbolic, using Lacan's expression. The symbolic is no longer able to provide the subject with identification tools; as such, the crisis of the symbolic pushes the subject to seek identification tools in technology. I will demonstrate this thesis through an analysis of AI narratives in technology journalism and scientific literature.

As I explained in the Introduction, there are many defition of AI. Generally, AI refers to "the study of agents that receive percepts from the environment and perform actions" (Russell and Norvig 2016, viii). Turner (2019, 16) defined AI as the "ability of a non-natural entity to make choices by an evaluative process." The dream of creating a machine capable of reproducing human intelligence is as old as humanity. However, starting in 2014, particularly after the acquisition of DeepMind by Google, the so-called machine learning revolution (Wooldridge 2021) began and brought unexpected and astonishing successes. The goal of machine learning is to establish programs that can compute the desired output from a given input without being given an explicit recipe for how to do this. Through training, the algorithms learn by themselves to recognize and extract useful patterns from data sets, and to connect inputs and outputs in the right way: "A machine learning algorithm takes a

DOI: 10.4324/9781003345572-8

data set as input and returns a model that encodes the patterns the algorithm extracted from the data" (Kelleher and Tierney 2018, 243). The explosion of so-called big data and new computational and statistical techniques related to machine learning (i.e., deep learning) have led to the evolution of increasingly complex systems that have achieved amazing results in many fields, including robotics, natural language processing, image recognition, automated translation, story writing, medicine, analytics, surveillance, and advertisement.

Overall, "AI can be said to refer to any computational system which can sense its environment, think, learn, and react in response (and cope with surprises) to such data-sensing" (Elliott 2021, 5). However, maybe the most important thing to say is that "AI is not so much an advancement of technology but rather *the metamorphosis of all technologies*" (Elliott 2021, 4; emphasis added). Like electricity, AI is an invisible force that governs the human world. The purpose of this chapter is to understand how this force interacts with the human unconscious mechanisms of identification. As tools for simulation and learning, AI-driven systems put into question our own ability to understand what human identity consists of. As Becker (2021, 107) writes,

> With artificial intelligence and robotics, engineering is no longer just about designing artifacts which imitate and exceed the ability of human workers. Nowadays, machines are more sensitive than ever before. They "see," "listen," "learn," "make decisions," and they tend to integrate more and more spaces once reserved to human beings. . . . Can one imagine that an artificial body, by simulating an inner activity or any sign of a subjectivity, can produce an effect of identity?

Psychoanalysis and the Deconstruction of Identity

Identity is not a central concept in Freud. The founder of psychoanalysis rarely speaks of identity or identification, yet psychoanalysis can teach a lot about identity. The very motto of psychoanalysis—*Wo Es war, soll Ich werden*—contains a fundamental indication: there is a clear difference between conscience, ego, and identity. If the first is essentially internal or external perception, while the second represents the psychic instance that mediates drives and social needs, identity is a much more complex and ambiguous process, and it involves the entire organization of the psyche. Each psychic instance (i.e., ego, superego, and id) has its own mechanisms of identification. Identity is therefore a process of identification, and the psychoanalytic work is a work on identification—a reconfiguration of damaged identification. In fact, psychoneuroses—that is, the object of psychoanalytic investigation—can be interpreted as pathologies of identification.

In Lesson 31 of the *Introduction to Psychoanalysis*, Freud distinguishes two general identification processes: (1) identification with the parental figure and (2) identification with the lost object, as occurs in processes such as mourning or melancholy. In another important book, *Group Psychology and the Analysis of the Ego*, we can distinguish at least four general types of identification, different but complementary:

1 *Primary*, or *pre-oedipal, identification*, which takes place in the presence of the object, is the fusional relationship between the infant and the maternal breast—in this phase, identification is incorporation.
2 *Secondary*, or *post-oedipal identification*, which takes place in the absence of the object, is the psychic substitute for an abandoned object bond—in essence, a defense mechanism to overcome the trauma of detachment from the mother's breast.
3 *Tertiary identification*, which has nothing to do with libidinal investment in the object, is the presence of common elements between subjects.
4 *Quaternary identification*, consisting of the acquisition of ideal aspects in which traits of the object can contribute to the constitution of the ego ideal, or the object itself can be put in the place of the ego ideal; this is a type of identification most visible in groups and masses.

A chapter would not be enough to adequately illustrate the theoretical problems associated with these types of identification. Is primary identification a phylogenetic or ontogenetic process? Is identification connected to incorporation, which in a certain sense seems to represent its somatic counterpart (Abraham 1924)? Is identification connected to introjection, which seems instead to constitute the psychic counterpart of incorporation? What is the difference, then, between incorporation, introjection, and identification, as Freud uses the terms *introjection* and *identification* quite interchangeably? Another problematic aspect concerns the relationship between identification and imitation: is imitation a primary and archaic form of identification?

The idea that identification is a complex and unstable process is also present in the work of Lacan. For Lacan, every identification is imaginary and, therefore, paranoid. The mirror stage and the infant's identification with the reflected image of themself are only the beginning of an illusion. In the mirror, the infant finds a unity of the self that is illusory, false. However, they believe in that image and then continue to project it on each subsequent object relationship. For this reason, the eruption of the symbolic, the social law, in the Oedipus complex represents a necessary stage in the psychic maturation of the subject. The symbolic breaks the cycle of imaginary and delusional identifications and gives the subject a new social identity (Lacan 2005).

The concept of projective identification has a crucial place in psychoanalytic theories on identity. Following the psychoanalysis of object relations inspired by Klein (1946) and Bion (1961), Grotsein (2009) defines *projective identification* as a strictly unconscious, omnipotent, and intrapsychic fantasy. He claims three theses: (1) there is no projection without identification, (2) it is necessary to distinguish projective identification from projective trans-identification, and (3) the normal and sublimated counterpart of projective identification is empathy.

Let me say something more about these three theses. Grotsein (2009) defines *projective identification* as mutual hypnosis and considers it to be a process acting in every stage of psychic life. In projective identification, A projects a part of themself—that is, psychic content (images, desire, fear, hate, etc.)—onto the psychic image of another subject, B. A modifies their own inner image of B by disidentifying a part of themself and transforming it into a part of B. An imaginative split of A's identity occurs. Grotsein argues that the psychic content disidentified and expelled from A nevertheless remains connected to A in the sense that it continues to identify itself with A. It is an even more complex unconscious process—the search for identification of the dis-identified by A. In other words, that content still bears the trace of A; it does not stop referring to A and giving information about A, as a sort of return of the repressed.

What happens to B, the target of the projection? According to Grotsein (2009), A not only modifies their inner image of B but also acts to influence B, that is, to modify their behavior in such a way as to make them like their psychic image. A intends to confirm their image of B by putting pressure on B. This is emphasized by Ogden (1982). The influencing behavior of A corresponds to an even more complex process in B; in fact, B is induced by A to unconsciously evoke those fantasies, emotions, desires, and thoughts that best correspond to the content projected and translated into action by A. Grotsein (2009) demonstrates this influence through numerous clinical examples. B operates in a state of semi-hypnotic submission, like a trance, which can lead to a projective counteridentification in the sense that B will, in turn, modify their internal image of A by responding to the pressure. In this case, recovering Bion's concept of "transformation into O," Grotsein speaks of "telepathy." There are telepathic abilities, extrasensory modes of communication, that act in the mechanism of projective identification. This means that the most elementary form of identification passes through a series of extrasensorial internalization and externalization processes in a purely psychic form of communication.

Projective identification takes place in the relationship between the mother and the infant. The latter, due to the pressure exerted by the accumulation of unbearable negative experiences in them, induces a symmetrical state of mind in the mother. The active pressure exerted by the infant arouses in the mother a corresponding emotional and imaginative state—this obviously only happens if the mother is a good mother, if she has a connection with the infant. The mother's psyche responds to the active pressure exerted by the infant by creating psychic contents similar to those of the infant. However, the mother has two possible responses: she can return to the child a pressure similar to their own, or she can return to the child a different pressure, one that shows the child how to accept those psychic contents that terrify them and how to instead give a positive response. A mother who is "good enough," to use Winnicott's expression, can respond positively to the infant by helping them manage their emotions and their relationship with the world. The failure of the mother's response, and therefore of the original identification process, is the origin of neurosis in the infant's life. Projective identification and counteridentification are unconscious social mechanisms that are constantly in place and that influence every human relationship with the world, as well as therapeutic practice in psychoanalysis.

This brief reconstruction of some psychoanalytic theories on identity highlights a decisive idea: in psychoanalysis, *there is no identification without disidentification*, and *there is no identity construction without identity deconstruction*. It is a paradox: dissociation is constitutive of identity. Psychoanalysis claims that we cannot think of identity as a substance or a solid core of the personality. Identification is an unstable, fragmented, fluid process made up of many subprocesses that are complex and dynamic. The root of psychoneurosis lies in the short-circuiting of this process. As Martini (2020) claims, delusion can be interpreted as a search for identity outside of social norms and natural evidence. The person who is delirious builds a fictional world that resists any denial of reality, to give themself an identity. In this way, they respond to the disintegration of their former identity. The delusional person must defend their delusions of themself; otherwise, they will fall into the total absence of identity, or in the inability to engage in a positive identification process.

Starting from these considerations, in the next section, I will show that a psychoanalysis of AI is, above all, an investigation into the unconscious tendencies that are at the root of AI. Why does the human being need to put intelligence, and even consciousness, into a machine? What are the desires and fears associated with this need for simulation and doppelgängers? How do they influence the identification of the human being? The rationalistic paradigm that monopolizes the current debate on AI does not give space or dignity to these questions.

AI and the Crisis of Identity in Contemporary Post-Fordist Capitalism

In its deconstruction of identity, psychoanalysis is very close to postmodernism. This is evident in Derrida and in his criticism of Lévi-Strauss. *Deconstruction* is a radical critique of presence and identity, understood as a stable essence that can be defined according to fixed categories. This line of thinking also emerges clearly in Butler's critique of sexual identity, as well as in Baudrillard's concept of simulation. Postmodernity consists of the clear rejection of the concepts of center, totality, and origin; the social space is fragmented by the multiplicity of linguistic games and by the crisis of the traditional mechanisms of legitimacy of political and scientific authority (Lyotar 1979).

In the last 20 years, sociological literature related to psychoanalysis has dealt with the theme of identity understood as a social practice—that is, the construction and deconstruction of the self through interaction with the social context. From this point of view, identity is primarily seen as a symbolic practice, involving self-image, biography, and one's vision of the future, values, work, love life, entertainment, social and institutional relations, politics, and so on. In the following, I intend to examine three key concepts in this literature: social acceleration, new individualism, and capitalist realism. My thesis is that these three concepts together define the crisis of human identification at the beginning of the twenty-first century. I claim that AI is, at the same time, the cause and the effect of this crisis.

Rosa (2013) provided a tripartite model of social acceleration. Post-Fordist capitalism has completely redesigned human time, making it more and more like the time of corporate processes, management, and digital technologies. Social and individual time has undergone an astonishing acceleration:

> My guiding hypothesis is that modernization is not only a multi-leveled process in time but also signifies first and foremost a structural (and culturally highly significant) transformation of time structures and horizons themselves. Accordingly, the direction of alteration is best captured by the concept of social acceleration.
>
> (Rosa 2013, 4)

Social acceleration is, therefore, a multifaceted process that acts on a multitude of micro and macro levels of social life and is influenced by multiple cultural, economic, technological, industrial, and political factors.

Rosa distinguishes three main forms of social acceleration: (1) technological acceleration, when there is a reduction in the amount of time it takes to achieve goals—oriented and intentional processes such as transport, communication, and production; (2) the acceleration of the pace of life, the scarcity of free time, and the pressure for a more productive lifestyle; and (3) the acceleration of society as a whole, when society's rate of change quickens so that there is a contraction of the amount of time it takes for social changes to occur. Hsu (2016) explores this tripartite scheme by analyzing individual forms of social acceleration. He distinguishes five individual forms of personal acceleration in the contemporary age: the detached self, the reflexive self, the reinventive self, the stationary self, and the decelerating self. Social acceleration turns out to be a contradictory phenomenon for the individual because it offers new possibilities and new restrictions at the same time, and therefore requires a continuous reinvention of identity.

This last aspect—the continuous reinvention of the self—is highlighted by Elliott (2014, 2015), who created the concept of "new individualism." The fundamental characteristics of new individualism are (1) continuous reinvention, (2) instant change, (3) speed, dynamism, and social acceleration, and (4) short-termism, or episodicity. New individualism is a social process in which actors (e.g., people, companies, and institutions) constantly reconfigure their identities. If, in past centuries, identity was based on the sense of belonging to a history, a society, or a culture, on being rooted in a place (e.g., a village, city, region, or state), today, it is based on transformation, plasticity, and plurality. "What is increasingly significant is how individuals recreate identities, the cultural forms through which people symbolize individual expression and desire, and perhaps above all, the speed with which identities can be reinvented and instantly transformed" (Elliott 2015, 51). New individualism has deep consequences on an emotional level, such as the development of anxieties and fears, the cancelation of traditional values and meanings, and the loss of a shared memory. The need for a constant reconfiguration or recreation of self-image, or self-history, can even fuel forms of depression—especially when social acceleration is not synonymous with change—whereby the individual becomes accustomed to continuous change and becomes inert, as Rosa (2013) claims through the concept of a "frenetic standstill."

Elliott's thesis is interesting because it differs from two other important theories of contemporary identity: the reflexive individualization based on Giddens' (1990) concept of "self-monitoring" and the notion of "technologies of the self" based on Foucault's concept of "self-surveillance." Unlike Giddens and Foucault, new individualism focuses on the imaginative contours of reinvention. The concept of reinvention is very close to that of reflexivity, as Giddens (1990, 38) describes: "The reflexivity of modern social life consists in the fact that social practices are constantly examined and reformed in the light of incoming information about those very practices, thus constitutively altering their character." However, Gidden's approach is still too rationalistic. For Elliott, self-invention is instead an unconscious imaginative and emotional process that responds to the precariousness and uncertainty of lives in globalized capitalism. From this point of view, Elliott's new individualism is much closer to that of Bauman (2000).

There are many examples of this contemporary need for self-makeover. The job market is the most classic. Companies are continually transforming their organization and goals, changing their locations and brands, and changing the roles of the people within them. This corresponds to the end of a stable and safe labor market: the worker must be ready to continually change duties, roles, skills, and sectors, with dramatic emotional consequences (Sennett 1998). "This is a corporate message that the self can be changed however the individual so desires: literally, there are no limits" (Elliott 2015, 59). Another good example is the phenomenon of "surgical culture," that is, the increase in cosmetic surgery and new forms of self-design, which is linked to Hollywood culture and the cult of movie stars and their bodies. The need for immediate solutions and results is closely connected with the constant and frenetic transformation of the body and image.

> In this new economy of short-term contracts, endless downsizings, just-in-time deliveries, and multiple careers, objective social transformations are mirrored at the level of everyday life. The demand for instant change, in other words, is widely perceived to demonstrate an appetite for—a willingness to embrace—change, flexibility, and adaptability.
>
> (Elliott 2015, 60)

I want to mention another element that belongs to the essence of post-Fordist capitalism, what Fisher (2009) calls "capitalist realism." As I mentioned before, Fisher's thesis is as simple as it is dramatic: Margaret Thatcher's assertion that "There is no alternative" to capitalism has been introjected at the level of the collective unconscious, with the result being that today, it is easier to imagine the end of the world than the end of capitalism. For us, human beings at the beginning of the twenty-first century, capitalism occupies the entire horizon of the desirable. There is an almost apocalyptic feeling in Fisher's words: capitalism has eroded the ideas, the sense of the future, and the hope of humanity. The power of capitalist realism "derives in part from the way that capitalism subsumes and consumes all of previous history: one effect of its 'system of equivalence' which can assign all cultural objects, whether they are religious iconography, pornography, or *Das Kapital*, a monetary value" (2009, 10). Capitalism is what is left "when beliefs have collapsed at the level of ritual or symbolic elaboration, and all that is left is the consumer-spectator, trudging through the ruins and the relics" (11). The main victims of capitalist realism are young people, whose hopes are torn apart by the imposed fatalism:

> What characterises capitalist realism is fatalism at the level of politics (where nothing much can ever change, except to move further in the direction of neoliberalisation) and magical voluntarism at the level of the individual: you can achieve anything, if you only do more training courses, listen to Mary Portas or Kirsty Allsopp, *try harder*.
>
> (Fisher 2018, 496)

Fisher underscores that capitalist realism is not a simple strategy of economic domination. It is also a type of esthetics, a form of entertainment and cultural manifestation. Capitalism presents itself as a necessary and sufficient pre-condition of culture:

> The point of capital's sponsorship of cultural and sporting events is not only the banal one of accruing brand awareness. Its more important function is to make it seem that capital's involvement is a precondition for culture as such. The presence of capitalist sigils on advertising for events forces a quasi-behaviouristic association, registered at the level of the nervous system more than of cognition, between capital and culture. It is a pervasive reinforcement of capitalist realism.
>
> (Fisher 2018, 516)

The consequences are evident—according to Fisher—in the excessive bureaucratization of societies, the diffusion of depression at the individual level, and the progressive destruction of the school and university system. Capitalist realism is the flattening of time to a single, monotonous present, characterized by feverish, small changes without a past or a future. On the one hand, the work of identification must be done more and more quickly; on the other hand, it remains nailed to a fixed and sterile present:

> When it actually arrives, capitalism brings with it a massive desacralization of culture. It is a system which is no longer governed by any transcendent Law; on the contrary, it dismantles all such codes, only to re-install them on an ad hoc basis. The limits of capitalism are not fixed by fiat but defined (and redefined) pragmatically and improvisationally. This makes capitalism very much like the Thing in John Carpenter's film of the same name: a monstrous, infinitely plastic entity, capable of metabolizing and absorbing anything with which it comes into contact.
>
> (Fisher 2009, 10)

Capitalist realism is both a belief (i.e., that "there is no alternative") and an attitude (defeatism, resignation, pessimism, and depression). Fisher's political judgment is that, before being a right-wing trend, capitalist realism is a pathology of the left:

> Really then, capitalist realism, whilst it is disseminated by the neoliberal right, and very successfully so, is a pathology of the left, or elements of the so-called left, that they succumb to. It was an attitude promoted by New Labour—what was New Labour if not instantiating the values of capitalist realism?
>
> (Fisher 2018, 524)

My thesis is that AI is closely connected to the crisis of identity described by the concepts of social acceleration, new individualism, and capitalist realism. From a psychoanalytic point of view, AI can be considered both the cause and the effect of a crisis of identification in the contemporary world. Now, my question is not whether or how AI transforms human identity. Turkle (1984), Hayles (2008), Fukuyama (2002), and other researchers have already clearly demonstrated the consequences of technological development on human psychology. From a psychoanalytic point of view, the real question is what drives the human being to seek their own identification in the machine; in other words, why does the human being need to build a machine designed and built to be like them, or even better than them, or to enhance their being through smart technologies? Where does the need for a posthuman identity come from? My thesis is that social acceleration, the need for a constant reconfiguration of identity, and capitalist realism made it necessary to "outsource" the work of identification to a machine. In other words, identification has become too complex and fast a process to be managed, even emotionally, by a single individual. As Braidotti points out, the posthuman subject of the contemporary world is "internally differentiated . . . not being framed by the ineluctable powers of significations, it is consequently not condemned to seek adequate representation of its existence with a system that is constitutionally incapable of granting due recognition" (2013, 188).

If we follow this line of thought, we must claim that social acceleration, new individualism, and capitalist realism are only three effects of a more original phenomenon in the collective unconscious: namely, the crisis of the symbolic order. In Lacanian terms, we are faced with an epochal laceration; the human subject can no longer rely on the symbolic to get out of the perennial circle of continuous pre-Oedipal imaginary identifications. Technology, then, emerges where the symbolic fails. The human subject entrusts their own search for identity to the machine when they no longer find space in the symbolic—that is, in language, society, law, or tradition—or when the symbolic is no longer able to give them their own space. The posthuman subject is a postsymbolic subject because they seek in technology what the symbolic can no longer provide—that is, an identity. Due to the crisis of the symbolic, the components of identity are no longer able to find harmony. Braidotti also underlines this point: "Lacan's notion of the symbolic is as outdated as a Polaroid shot of a world that has since moved on" (2013, 189).

From this point of view, the transformation of Lacan's interpretation from the 1970s and 1980s to the 1990s is also significant: after a hyperstructuralist reading, like that of Althusser, which exalts the symbolic to underline the inevitability of the social and political order as a structure of desire, a postmodern reading takes place, like that of Zizek, which instead exalts desire and the beginning of a post-political era. The latter

> seemed not only more in tune with contemporary cultural developments, but also offered a range of key theoretical concepts (from *objet petit a* to the notorious *mathemes*) that engaged multi-dimensional social realities in a fashion that other theoretical currents seemed unable to comprehend.
>
> (Elliott 2014, 4)

I do not want to discuss here the general validity of the Lacanian psychoanalytic model. I will just develop my thesis. As Elliott demonstrates, "there are two core premises of Lacan's Freud which should be rejected: first, that 'lack' transcendentally pierces and frames in advance the production of desire; second, that the conscious/unconscious dualism is best approached as a linguistic relation" (2014, 5).

As mentioned in the previous section, for Lacan, the symbolic was closely connected to the Oedipus complex: the subject interrupts the series of imaginary identifications—starting from the mirror stage—due to the intervention of the father and the social law. Lacan identified the father with the symbolic function. The Oedipus complex represents the structure of the sign, the original reference—that is, the original repression of incestuous desire. In the absence of the symbolic and the relationship with social order, the subject does not acquire the ability to manage desire and, therefore, continues to identify themself in a psychotic way. If positive identification depends on overcoming the Oedipus complex thanks to the symbolic, the subject without the symbolic is not a subject, has no identity, and is therefore forced to constantly reinvent themself in the grip of their ghosts and without reference to a shared reality.

From this point of view, technology represents the revolt of the imaginary against the symbolic. Following Castoriadis's (1975) work on the imagination, I claim that identification is always a work of the radical imagination, which cannot be reduced to the mere ability to replicate something (i.e., the imitative faculty). The imagination, for Castoriadis, is also radical in another sense: it is completely independent of language. In this way, Castoriadis invites us to "provincialize" human language and understand that semiosis, in a Peircean sense, goes far beyond language, that is, the human symbolic. One of the main theoretical issues of this book will be the development of a theory of the unconscious based not only on the concept of radical imagination but also on the Peircean conception of unlimited semiosis. The imagination has not only a negative function, therefore—as it has, for example, in the mirror stage in Lacan, where it is synonymous with paranoia, illusion, and delirium. Imagination also has a fully creative function, as the works of Kristeva and Laplanche also confirm.

The revolt of the imaginary against the symbolic is evident in technology such as augmented reality, which takes the real world and overlays it with computer-generated information or images. A good example of this is deepfake, an AI-based human image synthesis technique used to combine and overlay existing videos or images with original videos or images. Due to deepfake, it is possible to edit videos and photos at will to spread false news for political purposes (Wooldridge 2021, 297–98). Another good example is Google Glass, with which users can superimpose information or an image over everyday life, modifying it.

Object relations psychoanalysis can help us understand this phenomenon from another perspective. According to Bollas (1989), the entire human psychic experience is marked by the first relationship with the mother. The relationship with the mother is the relationship with what Bollas calls the "transformative object," an object that promises to transform the self and its world. The infant meets the mother not through a mental representation, or a desire for possession, but above all through an experience of radical change: the mother is a process of environmental and individual transformation. The research of the transformative object is subsequently moved by the infant toward other objects, such as Winnicott's transitional object, or toward symbolic objects, such as religious faith.

However, these movements always bear the mark of the first original relationship with the mother and, above all, with the way in which the mother was able to transform the child's experience in a positive or negative way. We can distinguish three moments in the research experience of the transformative object: (1) the search for the object—and, therefore, the investment, the hope, and the "psychic prayer" for the arrival of the "savior" object; (2) the fusion with the object, because the relationship with the transformative object is not mediated by a mental representation nor by the desire for the object itself but is a perceptive and existential experience; and (3) the memory of the ancient symbiotic relationship with the mother, a true original transformative object. Bollas clinically shows how being disappointed in the experience of the transformative object can lead to different psychopathologies—then, the analyst themself is identified with the transformative object. According to Bollas, the search for transformations of the transformative object is the most pervasive archaic object relationship and is a type of projective identification.

From this point of view, the relationship with AI can be interpreted as a search for the transformative object, a regressive movement to the maternal relationship in search of identification. A confirmation of this interpretation can come from various studies on the narrative of AI conveyed by the media. This research clearly shows the tendency of the collective imagination to see AI as an increasingly important agent in human society, something capable of radically changing the nature of this same society. An AI myth was born that obscures its reality and its dark sides.

Natale and Ballatore (2017) show how the rise of AI was accompanied by the construction of a powerful cultural myth: the creation of a thinking machine that would be able to perfectly simulate the cognitive faculties of the human mind. Based on a content analysis of AI-related articles published in two magazines, *Scientific American* and *New Scientist*, this research identifies three dominant patterns in the construction of the AI myth: (1) the recurrence of analogies and discursive shifts, by which ideas and concepts from other fields are employed to describe the functioning of AI technologies; (2) a rhetorical use of the future, imagining that present shortcomings and limitations will shortly be overcome; and (3) the relevance of controversies around the claims of AI, which they argue should be considered an integral part of the discourse surrounding the AI myth.

Bory (2019) compares two key events that marked the narratives around the emergence of AI in two different time frames: the game series between the Russian world champion Garry Kasparov and the IBM supercomputer Deep Blue held in New York in 1997 and the Go game series between the South Korean champion Lee Sedol and the Google DeepMind AI AlphaGo held in Seoul in 2016. Bory claims that these events represent a shift in the AI narrative. Interestingly, the paper investigates the way in which IBM and Google DeepMind used human–machine competition to narrate the emergence of a new, deeper form of AI. On the one hand, the Kasparov versus Deep Blue match was presented by the broadcasting media and by IBM itself as a conflictual and competitive form of struggle between humankind and a hardware-based, obscure, and humanlike player. On the other hand, the social and symbolic message promoted by DeepMind and the media conveyed a cooperative and fruitful interaction with a new software-based, transparent, and un-humanlike form of AI. This shift "reveals how AI companies mix narrative tropes, gaming, and spectacle in order to promote the newness and the main features of their products" (Bory 2019, 5). Recent narratives of AI based on human feelings and values, such as beauty and trust, "can shape the way in which the presence of intelligent systems is accepted and integrated in everyday life" (Bory 2019, 5).

Bourne (2019) and Goode (2018) show how the AI narrative was organized by large media and industry groups not only to promote a friendly and positive image of AI but also to highlight its importance for the future of humanity, often in a sensationalistic and misleading way. The dangers and defects of AI are not seriously discussed.

An interesting case is that of AI intuition, a new type of technology that has attracted growing interest from the specialized and journalistic press in recent years. Johanssen and Wang (2021) show that the interpretation of intuition prevalent in technology journalism websites was that of a rational procedure and reducible to algorithms. This is in complete contrast to the common conception of intuition; in the Western philosophical tradition, intuition is manly an unconscious ability, a form of immediate and emotional knowledge that varies from person to person. "While artificial intuition may appear humanlike, it cannot reach the complexity of human intuition as it is unique to each individual. Deploying intuition as a notion or model for AI is therefore problematic" (Johanssen and Wang 2021, 181). Commentators and journalists "misrecognize intuition as a technicality that can be added to AI in order to make it more flexible, dynamic, and autonomous" (Johanssen and Wang 2021, 181).

How can intuition be code-based? Is there not a distortion of human nature itself in this idea? Johanssen and Wang (2021) demonstrate that enthusiasm for artificial intuition blinded journalists and researchers not only to the reality of human nature but also to the potential dangers of this technology: "There is a danger in advocating intuitive AI because intuition is by definition difficult, if not impossible, to explain and account for. If AI acted truly intuitively, this could serve as a justification for being intransparent and opaque" (Johanssen and Wang 2021, 182). The result is that "AI is not only anthropomorphized and made more human through the advocacy for intuition; at the same time, humans are made more machinic, algorithmic, and technical than they really are in the data we examine" (Johanssen and Wang 2021, 183). In these works, there is "the desire to advance human subjectivity through AI" (Johanssen and Wang 2021, 184).

The same trend is also present in the specialized literature. Lovelock (2019) clearly states that AI is the only way to save not only humanity but also the Earth. Russell (2019) argues that AI "would represent a huge leap—a discontinuity—in our civilization. . . . [It is] the biggest event in human history" (2). Tegmark (2017) is no exception: "We're the guardians of the future of life now as we shape the age of AI" (335). Transhumanism, for example, in Kurzweil's formulation, is the reflective expression of this unconscious psychic research (Kurzweil 2005). Morozov's (2013) so-called techno-solutionism, which paints the world as full of bugs that can only be solved with the application of AI, also fits this line—the idea that AI will be able to guarantee access to information and knowledge for all, combat inequalities and poverty, and replace politics.

Humans, therefore, ask AI to transform their bodies and their world, and thus save their identity—that is, save them from the disintegration of their identity due to the crisis of the symbolic. Humans need new identification tools with which we can create and recreate new identities at a frenetic pace. Cinema confirms this expectation of existential transformation with films such as *I Am Mother* (2019), *Ex Machina* (2014), *AI Rising* (2018), and *Her* (2013).

The need to spread a positive image of AI is also a reaction to other deep fears, such as that of the double or that of a society in which humans become slaves to machines. As Becker (2021) writes,

> Robotics and artificial intelligence are perhaps even crossed by older anxieties, the fear of the figure of double, of objects that could come to life, the Promethean fear of exceeding the limits of the human

> experience. A dark magic seems at work when it comes to these machines, which can only be countered by the implementation of other techniques of enchantment *in order to allow humans to project on these artifacts a kind of trust that machines are yet unable to experience.* Artificial intelligence and robotics take shape between the dreams of entrepreneurs and the nightmare that we could one day live in a society of automatons.
>
> (108; emphasis added)

This research on AI narratives clearly shows that the emotional and imaginative investment in AI transcends the real conditions of AI and is the manifestation of the search for a transformation of human identity.

Humans are asking AI for a profound transformation of their world and of themselves; consequently, projective identification shifts from the subject to the object. The subject changes their psychic image of the object (e.g., the narrative transformation from Deep Blue to AlphaGo: the creation of the myth) and then exerts pressure on it to adapt it to that image and its need for transformation. In this process, the object is not neutral; it is able to respond, at least partially, to pressure. On the one hand, there is a "humanization of the algorithm," in the sense that humans entrust more and more human qualities to the machine, and the machine becomes a social agent who knows everything about their lives, even better than they do themselves. On the other hand, there is an "algorithmization of the human," in the sense that human beings tend to become more and more like the machine, to think of themselves as an algorithm (e.g., the idea of intuition as an algorithm, as discussed), and to entrust their future to the machine. In AI, "the human subject is regarded in a one-dimensional, functionalist way. Humans and AI function in the same way and are almost interchangeable. Such equations do not do justice to the complexity of human subjectivity. The human mind is more contradictory and messier than a computer" (Johanssen and Wang 2021, 183).

Therefore, in this section, I showed how the critique of identity in psychoanalysis can be extended to the contemporary age through sociology. I identified three key concepts to analyze this crisis: social acceleration, new individualism, and capitalist realism. I advanced the thesis that AI is both a cause and an effect of this process. In Lacanian terms, the crisis of identity in the world of post-Fordist capitalism is a crisis of the symbolic, which has become incapable of giving the individual an acceptable means of identification. The posthuman subject is a postsymbolic subject because it seeks in technology what the symbolic can no longer give it; that is, a strong identity. From the viewpoint of the psychoanalysis of object relations, the search for identification in AI can be interpreted as a type of projective identification, which Bollas called the desire for the transformative object. We ask the machine for a new identity capable of responding to our frailties.

I further showed in this section that an examination of the media narrative surrounding AI confirms this thesis. A paradox of identification emerges: the human being constructs their own identity by destroying it, in the sense that they project aspects of the self onto the machine, humanizing the machine. By doing this, they internalize aspects of the machine, getting used to thinking of themself as an algorithm and therefore betraying their true nature. There is no identification without disidentification.

The Anthropocene

The crisis of modern identity that is at the root of the complex phenomenon of AI is also connected to another phenomenon: the emergence of a new non-modern subject that is incomprehensible from the point of view of modern anthropology. This subject is Gaia, and it is the protagonist of a completely new era, the Anthropocene. Coined in 2000 by Paul Crutzen, a Nobel Prize winner in the field of chemistry, and indicating the supposed geological era after the Holocene in which humans become the main factor influencing the transformation of the environmental conditions of Earth, the term *Anthropocene* has known as an enormous success in the last 15 years, including in the fields of social science and philosophy (see Crutzen 2002, 2004; Angus 2016).

Far from being a geological epoch, the Anthropocene is the result of a "second Copernican revolution" (Angus 2016, 27), that is, a radical change in human relation with the Earth:

> Crucial to the emergence of this perspective has been the dawning awareness of two fundamental aspects of the nature of the planet. The first is that the Earth itself is a single system, within which the biosphere is an active, essential component. In terms of a sporting analogy, life is a player, not a spectator. Second, human activities are now so pervasive and profound in their consequences that they affect the Earth at a global scale in complex, interactive, and accelerating ways; humans now have the capacity to alter the Earth System in ways that threaten the very processes and components, both biotic and abiotic, upon which humans depend.
>
> (Steffen et al. 2005, 1)

At the very moment in which humans are attributed a destructive capacity toward the Earth, they become increasingly aware of how much humanity and its existence depend precisely on the Earth—that is, on nonhuman entities, such as atmospheric agents, technologies, other living beings, and the same geological layer on which they walk. This poses an ontological problem: who is this human subject who discovers that they are inseparable from the Earth and its integrated system? Furthermore, this is a moral problem; the causes of the impending catastrophe lie in a certain set of human activities and, therefore, in certain human groups. This moral responsibility also means that there is a need to identify new goals for humanity in and after the Anthropocene.

As Baranzoni et al. (2016, 7) suggest,

> the Anthropocene requires first of all to think of the possible end of human life on Earth caused by the human being himself, the radical problematic nature of a teleology of reason, that is, of an essentially human end inscribed on the horizon of humanity.
>
> (my translation)

From this perspective, the Anthropocene is the symptom of a triple unease: (1) that of the current Western capitalism, neuroticized by the absence of a future and by the anguish of its own economic castration; (2) that of the *mathesis universalis*, whose algorithmic rationality is deleting the differences that make up the *anthropos*; and (3) that of interdisciplinarity, as although the Anthropocene shows how humans cannot exist without otherness, a radical strategic ecology has still to be elaborated—that is, an ecological thinking capable of reflecting on the political and social relations between humans and their others.

The Anthropocene is closely connected with the idea of limits. As Guariento (2016) and Hamilton (2016) explain, the concept of the Anthropocene involves the awareness of the existence of non-negotiable limits to human action, or limits beyond which the existence of the human species would be jeopardized:

> The essential point of the Anthropocene concept is the opposite of the ecomodernist understanding. The Anthropocene is put forward not as a description of the further spread of human impacts on ecosystems but as a new epoch in the Geological Time Scale, a phase shift in the functioning of the Earth System. It is not a continuation of the past but a step change in the biogeological history of the Earth. The previous step change, out of the Pleistocene and into the Holocene, saw a 5°C change in global average temperature and a 120-m change in sea levels. Geologically speaking, the Anthropocene event, occurring over an extremely short period, has been a very abrupt regime shift, closer to an instance of catastrophism than uniformitarianism.
>
> (Hamilton 2016, 8)

The awareness of non-negotiable limits is the condition of the reconceptualization of the modern notion of nature. Haraway (1991), Latour (2017), Descola (2013), and Viveiros de Castro (2009) contest the mononaturalism that developed in Europe starting from the Copernican revolution, the mathematization of Galilean nature, the colonization of the Americas, and the transformation of the concept of sovereignty as a theological institution to protect against the emergence of the state of nature. This can be described as an "ontological turn": "The ontological-turn movement is an effort to take seriously different ontologies in different cultures (we have to bear in mind that knowing there are different ontologies and taking them seriously are two different things)" (Hui 2016, 2). The idea that nature is unique, passive, resilient, and constrained by a universal system of mathematical laws is contingent, both from a historical and an anthropological perspective. The Anthropocene requires us to reconceptualize our notions of the human being, human species, nature, and culture. "What is really new about perceptions of the Earth System over the last 10–15 years is the development of a perspective that embraces the Earth System as a whole. Several developments have led to this fundamental and accelerating change in scientific perception" (Steffen et al. 2005, 2).

Sloterdijk (2013) highlights how the concept of the Anthropocene implies a moral—juridical semantic nuance, as it emphasizes the human responsibility for the condition of the maintenance and stability of the entire biosphere, or rather, of the entire Earth System. In fact, if humans are considered effectively responsible for the development and health of the Earth and if, therefore, it is legitimate to bring a "trial" against humanity as a whole in relation to the good or bad administration of this responsibility, then it is necessary to consider the human being not as a merely biochemical mass but as a systemic power capable of influencing its environment. Therefore, the moral issue raised by the Anthropocene lies in the need to assess whether and how humans will be able to transform their environment for the better and ensure their survival.

The Anthropocene, for Sloterdijk (2013; Sloterdijk and Heinrichs 2001), is a consequence of globalization. This phenomenon is investigated throughout the *Spheres* trilogy and in *In the World Interior of Capital*. In *Spheres II* and *Spheres III*, Sloterdijk (2011–2016) distinguishes three forms of globalization. The first form of globalization, theological—metaphysical globalization, began with the first advanced cultures and involved the theorization of the terrestrial globe as a unique, perfect sphere capable of protecting the whole of humanity. The second form of globalization began with the great expeditions of conquest in the fifteenth century and culminated in the world order defined by World War II and by the Western model of capitalism. The third form, electronic globalization, is typical of the contemporary world and establishes an immediate and totalizing communicative system that instantly connects every part of the Earth, making it possible to effectively apply the hegemonic-capitalist dynamics of the world on a planetary scale. However, this system produces the definitive fragmentation of the totalized cosmos into a plurality of "foams," that is, spheres of action and conferment of meaning that are not completely autonomous or totally dependent on each other.

How do we respond to the ethical problem posed by the Anthropocene? According to Sloterdijk, the possibility of obtaining more with less—that is, of multiplying well-being while reducing the exploitation of resources—should be based on the adoption of homeotechnical practices instead of allotechnical practices. This pair of oppositional concepts was introduced by Sloterdijk for the first time in *The Domestication of Being*. Allotechnics practices represent the traditional Western way of thinking, which is based on the binary oppositions typical of classical metaphysics; according to this way of thinking, a spiritual and active human subject-master freely and indiscriminately imposes its will on a passive object-material servant. Instead, homeotechnical practices tend to *technically* prolong natural processes to establish or improve the collaboration between the different agents of the Earth System. The goal of homeotechnical practices is to make the agents cooperate for common advantages and not to create hierarchical relations of exploitation. This is the path indicated by Sloterdijk: technology must collaborate with the Earth. The conditions of a real homeotechnical practice are, on the one hand, the overcoming of the monovalent conception of being and the bivalent logic of Western metaphysics; on the other hand, they involve the development of intelligent technologies that would reshape the notions of object and matter. The collaboration between human and nonhuman agents could then lead, according to Sloterdijk, to the production of the Earth understood as the totality of the dynamics active in it.

This means, returning to what has been said in the previous sections of this chapter, that the construction and development of homeotechnical practices is the fundamental condition for satisfying the desire for identification of contemporary human beings. Such practices, however, will need to be based on a language common to humans and nonhumans. From this point of view, and this is my hypothesis, the reference to biosemiotics can be the starting point for building homeotechnical practices.

Conclusions

The AI revolution is not so much about the singularity or the supposed threat of cyborgs but about the daily life of human beings, their way of reflecting on themselves and building their own identity. In this chapter, I showed how the critique of identity in psychoanalysis provides important conceptual tools for understanding the AI revolution. In the first section, I proposed a thesis according to which AI is simultaneously the cause and effect of the identity crisis in post-Fordist capitalism. Three phenomena characterize our world: social acceleration, new individualism, and capitalist realism. These three phenomena can be interpreted as a global crisis of the symbolic, in Lacan's terms, and therefore as a global crisis of the traditional mechanisms of psychic identification. Digital technology, and in particular AI, is a response to this crisis. Humans seek AI not just as a tool but as a new form of identification.

7 Cybernetic Derrida

Différance and the Constitution of the Digital Object

The central thesis of the chapter is that to understand digital technology, it is necessary to develop the Derridean concept of *différance* in a different direction from that of Stiegler. In the first section, I will analyze the concept of différance in Derrida starting from Hägglund's interpretation. According to Hägglund, différance is not an ontological concept but a logical one. Différance is the becoming-space of time and the becoming-time of space; this co-implication produces the logical categories of succession and trace, and it characterizes life itself. In the second section, I will analyze the Stieglerian interpretation of différance. According to Stiegler, différance alone is not enough to define technology. Stiegler identifies technology and anthropogenesis; technology arises from a rupture in the history of the différance that corresponds to the appearance of the human being. In the third section, I will show how Stiegler's interpretation produces a merely functional definition of *technology*—it is the technology that contributes to the epiphylogenesis. Stiegler interprets the movement of différance as essentially homogenous, that is, as the simple repetition of the same mechanism genetically programmed until the rupture, represented by the human technical behavior.

Unlike Stiegler, Derrida does not think of différance in a homogeneous way. For Derrida, technology is not the effect of a rupture in life—that is, in différance—but as an emergence effect in the process of life itself. Thinking of the différance as anthropogenesis, and therefore as a rupture in the development of the différance, introduces a multiplication of the différance that has no reason to be. Following Derrida, I propose instead to identify technology with différance. This identification allows us (1) to obtain an ontological, and not merely functional, definition of technology and (2) to understand digital technology as an extremization of différance. Digital technology is not, in fact, a technology in the usual sense of the word; it is instead the extremization and transformation of all forms of technologies. From this point of view, in the last part of the chapter, I will develop an analysis of software. Software is a planetary infrastructure that today completely redefines the concepts of life and matter, of

DOI: 10.4324/9781003345572-9

human and nonhuman. To describe this infrastructure, I will draw on the concept of The Stack, coined by Benjamin Bratton. In the last part of the chapter, I will oppose The Stack to Gaia, as if they were two mythological figures in combat. Gaia is the unconscious of The Stack, the one that resists The Stack; Gaia is the great repressed that questions the hegemony of the digital. A philosophy of technology cannot fail to reflect on différance and, therefore, on the fight between The Stack and Gaia.

Interpreting Différance

The enigmatic term *différance* should not be understood as a concept or a word proper. It denotes something that does not exist—something that is not a being-present and that has no essence. Différance is pure movement,

> already plural and non-simple in its origin, for whose understanding our extant conceptual frames are ultimately inadequate. It simultaneously refers to both deferral in time and differentiation in space, constituting the originary interplay of time and space, the becoming-space of time (spatialization) and the becoming-time of space (temporalization).
>
> (Pavanini 2022, 4)

In this way, Derrida aims to deconstruct the oppositions engendered by western metaphysical tradition (i.e., form and matter, nature and culture, signified and signifier, etc.), ultimately questioning its conception of "being" as pure presence.

It would be wrong to understand the notion of différance only in linguistic or ontological terms. As Hägglund (2008, 2011) rightly suggests, différance is above all a logical notion. Indeed, Derrida defines *différance* in terms of a general co-implication of time and space; the term designates the becoming-space of time and the becoming-time of space, which Derrida abbreviates as *spacing* (*espacement*):

> The verb *différer* . . . has two meanings which seem quite distinct. . . . In this sense the Latin *differre* is not simply a translation of the Greek *diapherein*, . . . the distribution of meaning in the Greek *diapherein* does not comport one of the two motifs of the Latin *differre*, to wit, the action of putting off until later, of taking into account, of taking account of time and of the forces of an operation that implies an economical calculation, a detour, a delay, a relay, a reserve, a representation—concepts that I would summarize here in a word I have never used but that could be inscribed in this chain: temporization. *Différer* in this sense is to temporize, to take recourse, consciously or unconsciously, in the temporal and temporizing mediation of a detour that

> suspends the accomplishment or fulfillment of "desire" . . . this temporization is also temporalization and spacing, the becoming-time of space and the becoming-space of time. . . . The other sense of *différer* is the more common and identifiable one: to be not identical, to be other, discernible, etc. When dealing with differen(ts)(ds), a word that can be written with a final ts or a final ds, as you will, whether it is a question of dissimilar otherness or of allergic and polemical otherness, an interval, a spacing, must be produced between the other elements, and be produced with a certain perseverance in repetition.
> (Derrida 1982, 7–8)

Hägglund links this idea to the concept of succession: "Succession should here not be conflated with the chronology of linear time, but rather *accounts for the constitutive delay and deferral of any event*" (Hägglund 2011, 263; emphasis added). Without succession,

> nothing will have happened, whether retrospectively or prospectively, and Derrida analyzes this structure of the event in terms of a necessary spacing. Spacing is thus the condition for anything that is subject to succession, whether animate or inanimate, ideal, or material.
> (Hägglund 2011, 263)

Différance "is not an ontological stipulation but a logical structure that makes explicit what is implicit in the concept of succession;" emphasizing the logical status of the différance, also called trace, "does not mean to oppose it to ontology, epistemology, or phenomenology, but to insist that the trace is a metatheoretical notion that elucidates what is entailed by a commitment to succession in either of these registers" (Hägglund 2011, 263–264). The logical structure of différance "is expressive of any concept of succession—regardless of whether succession is understood in terms of an ontological, epistemological or phenomenological account of time" (Hägglund 2011, 264). The succession of time implies "that nothing ever is in itself; it is rather always already subjected to the alteration and destruction that is involved in ceasing-to-be. It follows that a temporal entity cannot be indivisible but depends on the structure of the trace" (Hägglund 2011, 269).

Différance, therefore, involves a radical critique of any concept of identity understood as a stable, fixed, absolute identity. However, a radical critique of identity must answer the question of the nature of identity itself: How can we have the perception of an identity? How can things be maintained despite the fragmentation of time? While Kant restricts time to a transcendental condition for the experience of a finite consciousness, Derrida claims (1) that there is no time and space as separate entities but only a constant process of the becoming-space of time and becoming-time of space, a process called *spacing*, and (2) that spacing "is an 'ultratranscendental' condition from which nothing can be exempt. The spacing of time is the condition not only for everything that can be cognized and experienced, but also for everything that can be thought and desired" (Hägglund 2008, 10).

This ultratrascendental is not a category or a supercategory of the subject because it conditions the subject itself and its identity. It is not a metaphysical principle because it is not stable. Instead, it is the affirmation of the pure instability of any principle; it is a radical thought of time that "dilutes" everything by dividing it into intervals. Nevertheless, Derrida distinguishes two moments: (1) the temporalization of space produces the deferral, the fragmentation of time and (2) the spatialization of time produces what Derrida calls the *trace*. The trace expresses the resistance of space to time, of simultaneity to succession; the trace allows for retention despite the fragmentation of time. Therefore, the trace is what remains of identity despite the fragmentation of time—despite différance. The spacing is the logical structure of whatever happens, of the event itself.

Hägglund's interpretation is interesting because it also extends to the world of biology and living systems. The structure of différance "is implicit not only in the temporality of the living but also in the disintegration of inanimate matter (e.g. the 'half-life' of isotopes)" (Hägglund 2011, 265). The logic of the succession can thereby "serve to elucidate philosophical stakes in the understanding of the relation between the living and the nonliving that has been handed down to us by modern science" (Hägglund 2011, 265). Therefore, the becoming-space of time and the becoming-time of space are the conditions that explain the disintegration of matter; the disintegration of matter responds to this double becoming. Living matter is matter that uses its organization to respond to the disintegration of matter, that is, to survive. According to Derrida, indeed, "mark, gramme, trace, and différance refer differentially to all living things, *all the relations between living and nonliving*" (Derrida 2008, 104; emphasis added).

From this point of view, Derrida's thought is a radical atheism, as it is a philosophy of survival and mortality that denies any kind of transcendence: "To survive is never to be absolutely present; it is to remain after a past that is no longer and to keep the memory of this past for a future that is not yet. . . . [E]very moment of life is a matter of survival, since it depends on what Derrida calls the structure of the trace" (Hägglund 2008, 1). This means that every moment—every *now*—

> passes away as soon as it comes to be and must therefore be inscribed as a trace to be at all. The trace enables the past to be retained, since it is characterized by the ability to remain despite temporal

> succession. The trace is thus the minimal condition for life to resist death in a movement of survival.
>
> (Hägglund 2008, 1)

The logic of survival is the logic of the concept of spacing; as Derrida points out in his late work *On Touching*, spacing is "the first word of any deconstruction, valid for space as well as time" (Derrida 2005, 207). *Deconstruction* is a philosophy of division and retention, according to which any claim to identity is illusory, and the task of deconstruction is to show this illusion.

> An interval must separate the present from what it is in order for the present to be itself, but this interval that constitutes it as present must, by the same token, divide the present in and of itself, thereby also dividing, along with the present, everything that is thought on the basis of the present.
>
> (Derrida 1982, 13)

The "Double Différance" in Stiegler

For Stiegler (1998), the birth of technology represents a rupture in the history of the différance. The essence of technology lies in a "double différance" (Stiegler 1998, 151), or a "différance of the différance" (Stiegler 1998, 177), a transformation of the différance that, according to Stiegler, Derrida underestimated or did not see.

The Stieglerian interpretation of différance emerges in *Technics and Time 1* in the context of a more general question—that of the "birth of the human" or of its "invention" (Stiegler 1998, 135). Stiegler deals with it through a critical reading of Leroi-Gourhan in which Derrida intervenes. "It is a question of identifying the evolutionary threshold that should allow us to distinguish the human from the animal, beyond the classic opposition between man and animal" (Vitale 2020, 1). My thesis is that by giving too much importance to anthropogenesis, Stiegler fails to grasp the logic of différance and—as Vitale (2020) also claims—re-proposes the hierarchically oriented oppositional structure that characterizes metaphysical thought and, in particular, the opposition between humans and animals.

According to Stiegler, the condition of human life is essentially technical. The technics make the human being, not the human being the technics. Anthropogenesis is a technical process and therefore—this is the essence of Stiegler's thesis—a new articulation of the différance. Technology involves a rupture in the history of the différance. If the différance is the disintegration of matter and, therefore, the need to survive that characterizes living life, then technics is nothing more than a new survival strategy. The process of anthropogenesis consists of the ability to reproduce the becoming-space of time and the becoming-time of space in the technical object and the process of epiphylogenesis—and therefore controlling this process, in some way. The différance is not the technology, but the technology transforms the différance in the sense that, as stated previously, it is a way to make the différance able to be experienced and thought about; it is this ability that distinguishes humans from animals.

Commenting on Derrida, Stiegler writes the following:

> Derrida bases his own thought of différance as a general history of life, that is, as a general history of the *gramme*, on the concept of program insofar as it can be found on both sides of such divides [*partitions*]. Since the *gramme* is older than the specifically human written forms, and because the letter is nothing without it, the conceptual unity that différance is contests the opposition animal/human and, in the same move, the opposition nature/culture. "Intentional consciousness" finds the origin of its possibility before the human; it is nothing else but "the emergence that makes the *gramme* appearing *as such*." *We are left with the question of determining what the conditions of such an emergence of the "gramme as such" are, and the consequences as to the general history of life and/or of the gramme. This will be our question.* The history of the *gramme* is that of electronic files and reading machines as well—a history of technics—which is the invention of the human. As object as well as subject. The technical inventing the human, the human inventing the technical. Technics as inventive as well as invented. This hypothesis destroys the traditional thought of technics, from Plato to Heidegger and beyond.
>
> (Stiegler 1998, 137)

Stiegler distinguishes two types of programs:

> on the one hand, the genetic program that would determine independently and exclusively the natural life of the living being in general and therefore also of the animal, to the point of determining its behavior as blind instinctive automatism. On the other hand, the cybernetic program with its various applications, technological product of conscious human action in a horizon that we should consider fully cultural, and that is independent of natural genetic determinism.
>
> (Vitale 2020, 3)

To understand anthropogenesis, it is therefore necessary—according to Stiegler—to add to the différance (i.e., the genetically determined one) a new, not genetically determined, différance:

> it would therefore be necessary to distinguish the différance at work in biological, natural, animal life from a différance that would be at work in the techno-cultural sphere proper to the human being, thus arriving at establishing the breaking point in the passage from the animal to the human being, from which to recognize another, radically new, determination of the différance.
>
> (Vitale 2020, 3)

In other words, the rupture from one différance to another is essential for Stiegler to understand the appearance of humans and the difference between humans and animals. This necessity is imposed by his interpretation of Lerroi-Gourhan's work.

The "double différance" is also the condition of the anticipation that is, for Stiegler, "the general and irreducible condition of the production of instruments and therefore as the manifestation of a technical intelligence that would already be creative" (Vitale 2020, 3). Indeed, Stiegler claims

> that it is starting from the epiphylogenetic trace, the trace that appears with technical life, that it is possible for us to discern the trace that constitutes life in general, and to access it, and not the other way around: this is a phenomenologico-existential standpoint in the strict sense, which makes conditions of appearance conditions of what appears. . . . To be able to access the trace that does not emerge from epiphylogenesis . . . it is necessary to start from epiphylogenesis, on the basis of epiphylogenesis. . . . Therefore, the trace before epiphylogenesis presents itself to us only through epiphylogenesis.
>
> (Stiegler 2020, 86)

Speaking of the rupture in the history of the différance and linking this rupture to anthropogenesis, Stiegler betrays the thought of Derrida, who instead sees a profound continuity between technology and human and animal life. Stiegler's thesis of the "double différance" is based

> on the possibility of denying, to the animal, to all the animals, a capacity of anticipation (and therefore also a mnemonic capacity) without which it is difficult to imagine its survival in a changing environment, rich in pitfalls and opportunities among which to orient oneself.
>
> (Vitale 2020, 3)

From this point of view, Stiegler confirms a metaphysical prejudice toward animals, functional to the definition of what would be properly and exclusively human: animal life determined by a blind instinctive automatism against the conscious freedom of human beings. This criticism is based on the biological notion of "genetic program" and the uncritical identification of the genetic program with Derrida's notion of différance.

Criticisms of Stiegler's theory of "double différance" have also been raised from the viewpoint of paleoanthropology and evolutionary biology. As Bentley-Condit and Smith (2010) claim, many nonhuman animals, and the great apes especially, produce and utilize technologies—however, their use and dissemination of these technologies are never systematic.

From Stiegler to Derrida

In this section, I intend to show how Stiegler's interpretation produces a simply functional definition of technology, in which technology is merely something that contributes to epiphylogenesis. Following Derrida, I propose instead to identify technology with différance and to rethink the concept of epiphylogenesis in a broader framework. This operation is motivated by two fundamental reasons: (1) I am convinced that the Derridean concept of différance allows us to formulate an ontological and not just a functional definition of technology and (2) digital technology—and in particular the notions of software and AI—requires us to overcome a purely anthropomorphic point of view on technology.

Stiegler interprets the movement of the différance at the heart of life as essentially homogenous, as the simple repetition of the same mechanism genetically programmed until the rupture, represented by the human technical behavior. *Différance*

> is the history of life in general, in which an articulation is produced (where art, artifice, the article of the name, and the article of death resonate), which is a stage of différance, and which had to be specified. The rupture is the passage from a genetic différance to a nongenetic différance, a "phusis differing and deferring."
>
> (Stiegler 1998, 175)

The prosthesis constitutes the human conscience, body, language, and sense of temporality and death:

> [the human] body and brain are defined by the existence of the tool, and they thereby become indissociable. It would be artificial to consider them separately, and it will therefore be necessary to study technics and its evolution just as one would study the evolution of living organisms. The technical object in its evolution is at once inorganic matter, inert, and organization of matter. The latter must operate according to the constraints to which organisms are submitted.
>
> (Stiegler 1998, 150)

Technics is an artificial memory that allows humans to anticipate, and therefore organize, experience. What differentiates humans from animal groups is that in the former the memory that holds the group together is external to individuals and not genetically determined, while in the latter, it is internal to individuals and genetically determined (Stiegler 1998, 155–157).

Therefore, Stiegler poses this fundamental equation: technics = externalization = temporal anticipation = anthropogenesis. Neanderthal man was born from an "instrumental maieutics" (Stiegler 1998, 158), that is, the co-determination of instrument and brain.

> In the process and in its evolution, the human undoubtedly remains the agent of differentiation, even though it is guided by the very thing it differentiates, even though it discovers itself and becomes differentiated in that process, in short, is invented or finds its image there, its imago, being here neither a phantasm nor a simulacrum—as it always is when describing technics.
>
> (Stiegler 1998, 158)

This process is unconscious, "analogous from this point of view to a zoological process. But the issue is not just one of analogy" (Stiegler 1998, 158). In instrumental maieutics, Stiegler identifies the "double différance"; "rather than being that of the human or the technical, the question is what absolutely unites them, time as the emergence of the 'gramme as such,' différance when it differs and defers in a new regime, a double différance" (Stiegler 1998, 151). Technics is a "double différance" because it is a "new organization of life" (Stiegler 1998, 164), of that life that is the différance, as such. Stiegler describes this "double différance" as " 'productive of difference,' as 'différance differed and deferred,' as a rupture *in* life in general qua différance, but not *with* life" (Stiegler 1998, 163).

Unlike Stiegler, Derrida does not think of différance in a homogeneous way. Derrida's logic of trace is not the repetition of the same but the opening of ever-new possibilities—of ever-new différances. In other terms, "différance refers to a process of differing/deferral, in which there is neither continuous homogeneity—pure repetition of the identical without difference—nor rupture—the emergence of the absolutely new, free from any vital conditioning" (Vitale 2020, 4). Therefore, technology is not the effect of a rupture in life—that is, in the différance—but an emergence effect in the process of life itself. Thinking of the différance as anthropogenesis, and therefore as a rupture in the development of the différance, introduces a multiplication of the différance that has no reason to exist. The risk in Stiegler's interpretation is that "the two 'moments' can be interpreted as the terms of a hierarchically oriented opposition: the opposition between vital différance (assumed as purely deterministic) and noetic différance (assumed as purely free)" (Vitale 2020, 4). This would mean returning to a classic metaphysical program.

By saying this, I do not mean that the concept of epiphylogenesis should be rejected. On the contrary, I think that it must be interpreted using a broader perspective. I propose to re-think the essence of technology, starting from the question of the animal in Derrida's work. Derrida (2008) shows the contradictions in western thought about the animal. He criticizes Heidegger especially. In comparison to *Dasein*, the animal is usually excluded from "being-towards-death" and, as such, does not properly die. And yet, the animal is accorded the character of a living being, in contrast to the inanimate or "worldless" stone. For Derrida, what distinguishes the human being from the animal is not the logos but *the nudity itself*. This is an essential point for Derrida (2008, 145). The animal is naked without knowing it, and therefore it is not naked because it does not know the concept of nudity. The human being is aware of their nakedness, and especially so in front of the animal, under the gaze of the animal, which is absolutely the Other. The relationship with the animal scares the human being not only because it reveals their nakedness; this relationship also scares the human being because in that nakedness, the human being discovers their animality—the Other appears closer, and the humans can discover the Other within themselves (see also Calarco 1972).

I claim that the same thing happens with technology. *Technology reveals the nakedness of the human being*, that is, the human's lack, or "poverty"—the différance as the essence of life itself, the disintegration of identity, and the need for iteration and survival. Technology is therefore not a rupture *in* life but its spontaneous development—the revelation of its logic. However, this development is not progress but a regression. Technology makes us return to the true nature of the human being. This is a fundamental characteristic of Derrida's thought: writing and technics do not speak so much of the future of humans—of their possible transhumanist enhancement—but of the past, of human history, and of the history of life in general as the history of survival.

Linking the concept of epiphylogenesis to that of différance, as Stiegler does, means engaging in an anthropomorphic interpretation of the Derridean concept, reducing its power and meaning. For instance, Stiegler states the following:

> How does grammatology pose this question? By calling man (or his unity) into question, and by forging the concept of différance, which is nothing else than the history of life. If grammatology thinks the graphie, and if in so doing it thinks the name of man, this is accomplished by elaborating a concept of différance that calls on the paleoanthropology of Leroi-Gourhan and does so to the extent that Leroi-Gourhan describes "the unity of man and the human adventure [no longer] by the simple possibility of the graphie in general, [but] rather as a stage or an articulation in the history of life—of

> what I have called différance—as the history of the gramme," while calling on the notion of program.
>
> (Stiegler 1998, 136)

In *Symbolic Misery 1*, Stiegler refers to "the tertiary retentions originarily constitutive of technical objects (inasmuch as they are epiphylogenetic)" (Stiegler 2014, 69). This means that "something is a technology only insofar as it partakes in epiphylogenesis, i.e., as it constitutes new organological articulations between biological organs, artificial organs and social organizations" (Pavanini 2022, 14). Stiegler then develops a functional definition of technology; *technology* is all that prolongs human individuation and, therefore, the effort to externalize memory. This means that we cannot say a priori what technology is and what it is not. It all depends on the contribution of that object to the epiphylogenesis process according to human needs. "Tertiary retentions are only active, however, and can only constitute this kind of support, on the condition that they are practiced" (Stiegler 2015, 107). This functional definition of technology betrays Derrida's thinking. For Derrida, *différance is technological and technology is différance*, in the sense that technology prolongs and strengthens the process of différance/survival.

Digital Différance

As Derrida writes, grammatology must be developed in terms of cybernetics:

> It [the différance] must of course be understood in the cybernetic sense, but cybernetics is itself intelligible only in terms of a history of the possibilities of the trace as the unity of a double movement of protention and retention. This movement goes far beyond the possibilities of 'intentional consciousness.' It is an emergence that makes the gramme appear as such (that is to say according to a new structure of non-presence) and undoubtedly makes possible the emergence of the systems of writing in the narrow sense.
>
> (Derrida 1974, 84)

Derrida is saying a few important things in this passage. The first is that the true meaning of différance is cybernetics; in his use of the term *cybernetics*, Derrida inevitably refers to the use of the same term in Wiener (1948) and, therefore, to the concept of automaton and the relationship between machines and living systems. The différance is cybernetic because it is the connection point between living systems and technological systems. Both living systems and technological systems must be understood in terms of the différance—this is their common source. The second important thing that Derrida is saying in this passage is that the logic of the trace exceeds the possibilities of intentional consciousness, that is, of the human subject. The logic of the trace is the condition of what Derrida calls "gramme" and which is "a new structure of non-presence." The *gramme* is the writing, the symbolic system. Writing is not a rupture in the logic of the trace but rather a manifestation "by emergence" of its development. For Derrida, writing is the origin of language and human consciousness, not the other way around.

However, the most important point of the passage we are analyzing is the third, which arises in the last part of the quotation: "In the narrow sense." By this, Derrida means that writing is only one of the possible manifestations of the logic of the trace. This leaves open the possibility of other manifestations of différance; différance evolves and transforms. Technology and design are other manifestations of this same logic. If we remain connected to an anthropomorphic perspective such as that of Stiegler, we fail to grasp this possibility; we fail to grasp the common origin of living and non-living systems and their connections. And above all, we fail to grasp how digital technology is an extremization of this same logic. Digital technology is an extremization of the logic of the trace because (1) it overcomes all forms of logocentrism and humanism and (2) it takes writing to its extremes.

My thesis is that Derrida's concept of différance gives us powerful conceptual resources to explain the core of digital technologies. The process of the datafication of experience—the reduction of each object to a set of data—is a manifestation of différance. The digital object is an object reduced to an unstable identity; every aspect of the object is a single phase of the succession. Data itself is nothing other than the single phase of an always-open succession. The acceleration of the becoming-space of time and of the becoming-time of space is what happens in our transistors; data is the result of this ongoing process. Data is a trace in the Derridean sense; it is unstable because it is a deferral that must replicate the fragmentation of time.

The datafication of experience is the condition of a further form of différance, that which occurs in software. Each object in the digital world can be reduced to a set of lines of code, in other words, software (Frabetti 2014; Possati 2022).

> The transformation of objects into signs has been greatly accelerated by the spread of computers. It is obvious that digitalization has done a lot to expand semiotics to the core of objectivity: when almost every feature of digitalized artefacts is 'written down' in codes and software, it is no wonder that hermeneutics have seeped deeper and deeper into the very definition of materiality.
>
> (Latour 2008, 4)

Two characteristics of software must be underlined here:

1. Software is writing, and yet it is by no means a form of writing like any other. It is writing meant solely to be written—there is no software orality; "reading" a code means re-writing it elsewhere (Chun 2011, 45).
2. Software is writing made to be executed, that is, to be materialized. Any digital object—an image on a screen, a music program, a video game, and so on—is the materialization of a code. Furthermore, the execution of the code is not human at all.

Software therefore questions the classic dichotomies of the western philosophical tradition, such as the material/virtual and universal/singular dichotomies. Software is a ubiquitous entity; it can run on many computers in different places around the world at the same time. Furthermore, software is a singular universal in the sense that every copy of a software is always the same software—there is no reference to an initial model—and, at the same time, it is something different that acts in different ways and modifies the environment in different ways—think, for instance, of AI machine learning systems.

Software is, then, the last phase of "an exteriorization always already begun but always larger than the trace which, beginning from the elementary programmes of so-called 'instinctive' behavior up to the constitution of electronic card indexes and reading machines, *enlarges différance and the possibility of putting in reserve*" (Derrida 1974, 84; emphasis added). The différance as software overcomes the différance as writing because it extends to everything—inanimate matter and animate matter, to redefine both. This means that digital technology has a planetary vocation. Bratton (2015) shows how we should not think of software simply as the action of a lonely nerd locked in their room but as a planetary infrastructure that today regulates and redefines everything, including human identity. In the rest of this section, I want to explore this new concept, The Stack, introduced by Bratton as an expression of digital différance. Furthermore, I propose to relate The Stack with another concept, which in a certain sense represents its "dark side," its antagonist: Gaia.

The Stack

Bratton (2015) identifies six layers of The Stack: Earth, Cloud, City, Address, Interface, and User. *Earth* entails the material and energy-harnessing geological demands of computing; *Cloud* names the weird sovereignty of corporatized, global technology services such as Google; *City* relates to the lived experience of cloud-computerized daily life; *Address* deals with identification as a form of management and control; *Interface* deals with coupling users to computers; and *User* relates to the human and nonhuman agents that interact with computational machines.

The appearance of The Stack is the expression and consequence of a radical change of era. This planetary architecture that defines almost everything is based on computation and imposes new forms of geography, politics, and sovereignty. In an article for the journal *Noema*, Bratton states that the appearance of this computational architecture should be understood as an evolution of terrestrial life, not as the opposite:

> The emergence of *planetary-scale computation* thus appears as both a geological and geophilosophical fact. In addition to evolving countless animal, vegetal and microbial species, Earth has also very recently evolved a smart exoskeleton, a distributed sensory organ and cognitive layer capable of calculating things like: How old is the planet? Is the planet getting warmer? The knowledge of "climate change" is an epistemological accomplishment of planetary-scale computation.
> (Bratton 2021, 1)

The Stack is a geological and biochemical effect. Over the past few centuries,

> humans have chaotically and in many cases accidentally transformed Earth's ecosystems. Now, in response, the emergent intelligence represented by planetary-scale computation makes it possible, and indeed necessary, to conceive an intentional, directed, and worthwhile planetary-scale *terraforming*.
> (Bratton 2021, 1)

The Stack is a "multilayered structure of software, hardware, and network stacks that arrange different technologies vertically within a modular, interdependent order" (Bratton 2015, 25). The Stack is a new Leviathan that is not composed of humans but of lines of code that follow "a general logic of platforms" (Bratton 2015, 25) incompatible with that of the twentieth century and modernity. "Maps of horizontal global space can't account for all the overlapping layers that create a thickened vertical jurisdictional complexity, or for how we already use them to design and govern our worlds" (Bratton 2015, 26).

In The Stack, technology is an extension not only of the human but also of the planet and its geography—The Stack asks that we learn to see "the designability of geography in relation to the designability of computation" (Bratton 2015, 30). Planetary-scale computation takes different forms at different scales: energy and mineral sourcing and grids; subterranean cloud infrastructure; urban software and public service privatization; massive universal addressing systems; interfaces drawn by the augmentation of the hand or the eye, or dissolved into objects;

sophisticated AI systems; and users both over-outlined by self-quantification and exploded by the arrival of legions of sensors, algorithms, and robots. The Stack therefore does not have a fixed, stable identity; it transforms according to the geographical and geopolitical conditions it finds in order to modify them according to its objectives, abandoning the traditional modern forms of state and power, as well as the traditional forms of biological life.

> The Stack both does and does not exist as such; it is both an idea and a thing; it is a machine that serves as a schema as much as it is a schema of machines. . . . The Stack is simultaneously a portrait of the system we have but perhaps do not recognize, and an antecedent of a future territory.
>
> (Bratton 2015, 26–27)

Nonetheless, The Stack maintains a unity, a coherence. "Instead of seeing all of these as a hodgepodge of different species of computing, spinning out on their own at different scales and tempos, we should see them as forming a coherent and interdependent whole" (Bratton 2015, 26). This machine, which is also an apparatus of distributed, diffused power, is not the result of a global project or a revolution. Bratton defines *The Stack* as "an accidental megastructure" (2015, 32) that is the result of successes and failures, of human projects but also of chances. A concrete example is Google, which is an actor of The Stack. Google is not simply a network of interacting human and nonhuman actants; it is a completely new entity, a new form of agent, to understand which the symmetric ontology of ANT is not enough. Google is an agent that operates according to a geopolitical logic that comes into conflict with traditional states and their institutions. It is a "nonstate actor operating with the force of a state, but unlike modern states, it is not defined by a single specific territorial contiguity" (Bratton 2015, 35). Google is a US-headquartered corporation but also a transnational actor that has taken on many traditional functions of nation-states. "While Google is as reliant on real physical infrastructure—its data centers are by no means virtual—that physicality is more dispersed and distributed than partitioned and circumscribed" (Bratton 2015, 35).

I propose to view the six layers that make up The Stack as six levels of a planetary individuation process. The Stack can be interpreted as a new type of formation of human and nonhuman identity. Each entity in The Stack is defined by the interaction of the six levels. The User, human or nonhuman, is defined by the relationship between the Interface and the Address and, therefore, by the set of material and virtual infrastructures that make this relationship possible—namely, the City and the Cloud. The Interface and Address configure the place of the individual in the structure and, therefore, the set of terrestrial resources it is possible to access. For example, the integrated design of driverless cars "includes navigation interfaces, computationally intensive and environmentally aware rolling hardware, and street systems that can stage the network effects of hundreds of thousands speeding robots at once. The next stable form of the 'automobile' may be as a mobile Cloud platform inside of which Users navigate the City layer of a larger Stack according to augmented scenery Interfacial overlays and powered by grids of electrons as well as bits" (Bratton 2015, 37). The unity of The Stack is therefore a complex unity that is based on a triple logic: (1) the overlapping of the layers, (2) the balance between the layers, and (3) the emergence. The ongoing intersection of these three logics constitutes the design of The Stack. This is also the epiphylogenesis in The Stack, a process of constant design and redesign that goes through all the stages of The Stack.

Gaia

I propose to distinguish two levels in the concept of Gaia: an ontological level and a normative level. The first level is "the intrusion of Gaia" (Stengers 2015), related to the ecological crisis. As Latour (2017) points out, Gaia is an agent that was inconceivable in the past. It is a completely new form of agency; it is neither the environment nor the nature of moderns—that is, the immutable nature conceived as an unalterable background of human affairs. Gaia is a historical agent that has been wounded and "moved," as Serres (1995) writes, and so rebels against human actions. What was considered the objective par excellence is animated.

Gaia is the name that the British inventor James Lovelock and the American biologist Lynn Margulis gave to our planet at the beginning of the seventies to indicate one characteristic: being alive, in the sense that this planet cannot be understood except as life; the presence of life shapes everything else (Lovelock 2000; Lovelock and Margulis 1974). Gaia is a complex being who arises from the interaction between life and inorganic beings—a fragile but necessary balance (on the debate on the Gaia hypothesis, see Schneider et al. 2004). Gaia arises as a social actor from the ecological problems posed by the Anthropocene and that—as Latour (2017) points out—modernity prevented from being fully addressed. "The drama—Latour claims—is that the intrusion of Gaia is happening at a moment when the figure of the human has never appeared so ill-adapted to take it into account" (Latour 2017, 107) because of the economic monoculture of capitalism.

As Stengers (2015, 43–44) writes, Gaia must be recognized as a living being and not just a sum of physical processes. Gaia is a being with a history and whose processes are all interconnected in a wholeness. "To question Gaia then is to question something that holds together in its own particular manner, and the questions that are addressed to any of its constituent processes can bring into play a sometimes-unexpected response involving them all" (Stengers 2015, 45). Stengers stresses that today, because of the climatic crisis, our understanding of the manner in which Gaia holds together is much less reassuring: "The question posed by the growing concentration of so-called greenhouse gases is provoking a cascading set of responses that scientists are only just starting to identify" (Stengers 2015, 45).

Is Gaia another Leviathan? No. Gaia is the demonstration that any sort of thinking that ignores the material and climatic "envelopes" that make its existence possible is pure madness. Gaia is an active being who reacts against threats. The ecological crisis makes human responsibilities evident and asks for a change of behavior. However, there is also something more: Gaia itself is a process, a set of contingent processes that have made some events more probable than others (Latour 2017, 45). Every little part of this process is active, participating in the process and its evolution. Each small part is responsible and is called to act in a certain way. Gaia is therefore also a great moral force because it evokes a form of micro- and macro-responsibility.

I see Gaia as the internal and external resistance to The Stack. It is internal because the processes of Gaia move within The Stack; they condition and are conditioned by the planetary software infrastructure. It is external because Gaia is the ultimate envelope, within which The Stack must remain to thrive. Gaia can be interpreted starting from the Derridean concept of self-immunity: that which protects, defends, repairs, saves from death, from extinction, and at the same time threatens with death what it protects (Derrida 1993). This means that life cannot be separated from death and that death is never beyond life, as something outside of it. These are the two faces of Gaia.

Conclusions

In this chapter, I developed an interpretation of the Derridean concept of différance in relation to technology. I therefore criticized Stiegler's interpretation of différance. I showed how Derrida's concept of différance offers us a definition of technology that is not merely functional. I also showed how the concept of différance is particularly effective in explaining digital technology. Digital identity is a form of différance; data is a trace, a pure differing that responds to the fragmentation of time without constituting itself in a solid, absolute, defined identity. This same instability is at work in the relation between The Stack and Gaia. The image of the fight between The Stack and Gaia particularly fascinated me; I find it to be the fundamental confrontation of the Anthropocene, in which what is at stake is the future of humanity—the human project for the twenty-first century.

Conclusions

A Planetary Negotiation

In "Nosedive," the first episode of the third season of *Black Mirror*, Lacie, the protagonist, develops an obsession with a social classification system, which not only evaluates and publicizes her social status (not just to friends and contacts, but everyone) but also, more generally, provides an *ad personam* assessment for government institutions and service industries. The episode shows how deeply all this affects Lacie's reputation and opportunities and provides an analysis of the deleterious effects that this policy could have on the sanity of citizens, the very citizens that the political authorities claim to protect.

The value of *Black Mirror* lies in its examination of the effects, not the causes. Each episode portrays a future society in which powerful technologies have been developed, but the social body has not yet absorbed these changes as well as it should have. It is this singular asymmetry that makes this series not only a fascinating dystopia of the post-Anthropocene world but also a reflection on the place of humanity at the beginning of the twenty-first century. The series offers imaginative variations that improve our understanding of the implications of technological development. This is achieved by focusing not on what these technologies are, but on the consequences that they could have on individuals and society. Specifically, these variations show the consequences of digitizing content and making it virtual, creating boundless connected networks, and administering huge quantities of data. *Black Mirror*'s method is to lead us to the threshold of the present and push us just one step further. Most of the episodes feature an unusual use of various technologies, either completely unthought of or currently imaginary, although some contain a reference to a previous use (e.g., the blocking of the visual and auditory abilities of a single subject or group in the episode "White Christmas"). The effects of these technologies are as disconcerting as they are fascinating, and some episodes may require several views to really understand the dynamics.

In this book, I intended to adopt the same approach as the series: to show how technologies today have reached such pervasiveness that they are able to shape human identity and even the unconscious. However,

DOI: 10.4324/9781003345572-10

that is not all. I maintain that the unconscious itself is a black mirror into which we are forced to look, an already shattered mirror and that it is the task of psychoanalysis to try to recompose what we see. As Latour writes:

> From this point on, the past has an altered form, since it is no more archaic that what lies ahead. *As for the future, it has been shattered to bits*. We shall no longer be able to emancipate ourselves the way we could before. An entirely new situation: behind us, attachments; ahead of us, ever more attachments. Suspension of the "modernization front." End of emancipation as the only possible destiny. And what is worse: "we" no longer know *who* we are, nor of course where we are, we who had believed we were modern . . . End of modernization. End of story. Time to start over.
>
> (Latour 2013, 10; emphasis mine)

Technoanalysis is an extension of psychoanalysis through (a) biosemiotics, (b) ANT, and (c) MET. It is therefore a multifaceted approach to the study of technology. The central thesis of technoanalysis is that the mind is essentially technical in the sense that it is a phenomenon emerging from the interaction between (a) natural semiosis, (b) material things, and (c) epiphylogenesis. The mind is therefore a web of human and nonhuman actants in constant interaction. Consciousness and the unconscious are properties distributed in the actor network. In other words, artifacts actively shape our instinctual world, generating the distinction between consciousness and unconscious. The technology is coextensive with cognition and consciousness, and therefore with the unconscious itself.

This book was, above all, an attempt at a theoretical reevaluation of psychoanalysis, starting from the Freudian technique. The thesis I advanced in Chapter 1 is that Freudian psychoanalysis is not only the single true form of scientific psychology (i.e., because it solves the problem of suggestion through the MES), but it is also a natural science based on a biosemiotic approach to the study of the mind. Therefore, I have shown how the Freudian paradigm can be integrated and strengthened through research on biosemiotics.

The second step in the theoretical reassessment of psychoanalysis in this book concerned ANT. Can we deploy Freudian psychoanalysis outside the modern subject/object dualisms? This question prompted us to radically reformulate the problem of the mind. Freud, Latour, and biosemiotics lead us to think of the mind as a hybrid comprising humans and nonhumans as a network of constantly interacting actants. Consciousness and the unconscious become properties distributed among the different actants of the network. The contribution of the MET has allowed us to formulate this initial intuition with greater precision: the reel mentioned by Freud in *Beyond the Pleasure Principle* is unconscious in the sense that it constitutes the unconscious of the child by shaping the drive and orienting it. The reel is not an inert extension of the child's unconscious. Instead, it is an active part of it. The reel materializes the drive and connects it to a whole series of things and meanings external to the child's brain. Based on this idea, I assume the "methodological fetishism" proposed by Appadurai (1986, 5) and Malafouris (2013, 133–136). As Ellen (1988) argues, fetishism is not the manifestation of an archaic mentality, but a fundamental characteristic of material engagement.

What remains of Freudian psychoanalysis in our research? Why did we start with Freudian psychoanalysis? First, I decided to adopt a psychoanalytic point of view because—as I argued in the introduction of this book—psychoanalysis is the only true science of the mind. In fact, psychoanalysis is the only form of psychology that solves the problem of suggestion. This does not automatically mean that all the results of Freud's work must be accepted as they are. I have shown that the Freudian method spontaneously extends in the direction of biosemiotics and ANT. The psychoanalytic mind is biosemiotic and organized in associations, in the sense that "it designates a *series of associations* revealed thanks to a trial—consisting in the surprises of the ethnographic investigation—that makes it possible to understand through what series of small discontinuities it is appropriate to pass to obtain a certain continuity of action" (Latour 2013, 33). This "principle of free association—or, to put it more precisely, this principle of irreduction—that is found at the heart of the actor-network theory" (Latour 2013, 33) is also shared by psychoanalysis. In light of this union between biosemiotics, psychoanalysis, and ANT, we are forced to reconceptualize the main notions of psychoanalysis, starting with repression and the unconscious. This is what I have done in this book by introducing the concept of anti-mediation which is connected to the postphenomenological notion of mediation and the immunization theory. I have shown that a specific form of anti-mediation is interaction failure in social robotics and how the Freudian method (interpretation + construction through the MES) can be applied to the study of it.

The repressed is therefore the name of a part of the network-collective, that is, the points of resistance in what Latour and Callon call the translation process in the network. The unconscious consists of things, ideas, and people that resist translation and force other things, ideas, and people to continually deviate and transform. Consciousness, in contrast, corresponds to the stabilization of the network and the establishment of constant, fixed translation relations. Accordingly, this thesis can be thought of as a radicalization of the MET and Stiegler's philosophy of technology. It is a radicalization because it is not limited to technology; the human being, according to my thesis, is thinkable only in relation to material culture. Technology is nothing more than a reflection of this co-constituting relationship.

This idea is the basis of the methodology, which I have called technoanalysis. In combining ANT and the post-phenomenological theory of mediation, technoanalysis examines the forms of technological mediation in the networks of humans and nonhumans to identify the anti-mediations internal to these networks. Technoanalysis is at the same time a form of reverse engineering (i.e., it moves from the technological object to its description) and of interactive design (i.e., it proposes solutions to "cure" anti-mediations). Anti-mediation is not a malfunction, although a malfunction may be an indication of anti-mediation. Anti-mediation is instead a phenomenological concept; it presupposes the concepts of epiphylogenesis and material engagement. One of the examples I have provided is that of interaction failure in social robotics. The concept of interaction failure is very complex because it is located on the border between technics and phenomenology.

The case of Replika is also an example of anti-mediation. Although designed as a psychological support mechanism, this AI chatbot has been accused of inciting murder. The study of technological mediation, in this case, passes through an analysis of the way in which the designers of Replika have told of its birth and development. I showed how Replika is a part of the unconscious of its designers by prolonging and transforming a mourning experience, which was, in this case, the loss of a friend. I pointed out the processes of externalization and anonymization of the unconscious that take place here. This shows that the concept of anti-mediation is coherent and can be the subject of systematic investigation.

The purpose of technoanalysis, therefore, is not only to make the technology visible, but also to show the set of processes that lead to the establishment and stabilization of a network. The purpose of technoanalysis is to investigate what it is in that set of processes that threatens the stability of the network (i.e., what we have called anti-mediation). Technoanalysis allows us to interpret these threats based on biosemiotics and the idea of the biological origin of technology. If the psyche is a bio-collective, and technology is therefore an integral part of it, the goal of technoanalysis is to understand the connection between the biological unconscious and its technology—or how the unconscious "speaks" in technology and technology in the unconscious. The basic inspiration of technoanalysis is Freudian because the unconscious is understood primarily as a form of repression and resistance. However, technoanalysis does not encourage us to mechanically apply Freudian theories to technology and AI. Technoanalysis takes MET very seriously and extends it to the emotional sphere. This means that Freudian concepts must be reconceptualized, starting from the analysis of technological mediations and anti-mediations. *Technological mediations and anti-mediations define the intimate history of the human being*. Anti-mediation *is* the repressed, and not a secondary form of the repressed. We must therefore take the opposite path: not from the psyche to technology, but from technology to the psyche. This is why the connection between technology, biosemiotics, and organology is so important for technoanalysis.

The last two chapters of the book are the most philosophical. In them, I introduced the theme of the Anthropocene, which is crucial to my analysis. We can no longer continue to have the same mental coordinate system that we had in the twentieth century. We can no longer adhere to the same understanding of knowledge that we had just 40 years ago. The concept of the Anthropocene raises the problem of the limits of technological progress based on the late capitalist model. This means that we must reconceptualize technology according to the new horizon of Gaia. From this perspective, I maintain that the fight between The Stack and Gaia is the most representative way to depict this situation. As philosophers, sociologists, and anthropologists, our duty is to negotiate and find a diplomatic solution for the sake of peace—in other words, a "planetary negotiation" (Latour 2013, 17).

There is one essential thread linking the concept of anti-mediation to the concept of Gaia. Both represent something that hinders technology *in technology itself*. This is something that, while being part of the technological system, challenges and puts into crisis the ability of that system to mediate between humans and nonhumans, as well as between nonhumans themselves. Thinking about technology today means considering the conflict between The Stack and Gaia and contemplating possible diplomatic solutions. As psychoanalysis teaches, defeating a resistance does not mean destroying it; otherwise, that resistance will come back stronger than before. Defeating resistance means recognizing its value and giving it the word. *Let Gaia speak through The Stack*; this must be the diplomatic effort put forth—a true homeotechnical practice, as Sloterdijk would say.

This research is part of a much broader project that investigates the modes of digital existence in the Anthropocene. The goal is to develop a new system of coordinates to improve the way we control and understand technology. I propose technoanalysis as a useful method and conceptual apparatus for integrating the different types of mediations and improving our understanding of the relations between humans and nonhumans. I cannot say that I have all the answers. I hope that I have demonstrated that we can at least begin to ask the questions.

Epilogue

In the last scene of *2001: A Space Odyssey*, Bowman suddenly finds himself alone in a room furnished in a Baroque style. Undecipherable sounds in the distance can be heard. Then, a mirror appears right above a bathtub. The mirror reveals the protagonist's aging and the change in his state. Bowman thus becomes aware of his sudden alteration in physical appearance, and his gaze becomes filled with disbelief and terror. Soon after, Bowman finds himself in the bedroom. He is sitting at the table and eating quietly. The transition from the mirror, the stage of imagined identification, to the bedroom, the stage of intrauterine and non-imaginary identification, is established by the breaking of the glass, which Bowman inadvertently bumps into, causing it to fall. It is the breaking of the glass that frees Bowman from the imaginary identification, or the cycle of paranoid identifications (Lacan 2005), and causes him to return to the uterine duality, or to a form of identification that is a carnal fusion with another human being (Sloterdijk 2011–2016). A few seconds after the breaking of the glass, Bowman falls ill, and he nears his life's end, causing him to lie in the bed. The bed symbolizes the fetal bond or the first intimate space (Sloterdijk 2011–2016, 337). Death is a return to the origin (i.e., the womb) and an intimate connection with a mother. The future and past coincide here; the spherical shape behind Bowman in the headboard of the bed is similar in form to a uterus, the circumscription of the womb, which welcomes him back into its walls. Death is the future and birth is the past, and at this moment of death, the past and future overlap.

The transition from the bathroom to the bedroom is perhaps the most paradoxical moment in the movie. Bowman experiences an accelerated future through the sudden aging process that allows him to return to the past, specifically the womb. At the moment of Bowman's death, technology appears again, as represented by the monolith. The unity of the monolith and the uterus is realized by the final placenta that appears on the bed. Bowman returned to his mother, and he once again becomes a baby protected by his placenta. The placenta is the first nobject, according to Sloterdijk (2011–2016, 339). It is the first form of human individuation, the original human "double" devalued by modernity (Sloterdijk 2011–2016, 335). Bowman's regression to his attachment to the placenta is similar to that experienced by HAL 9000 shortly prior in the movie. Bowman finds himself in the same situation as HAL 9000. Therefore, the regression of technology hints at the possibility of human regression. In technological regression—or rather, in the failure of technology—HAL 9000 has rebelled. Thus, regression takes here the form of anti-mediation; the human being experiences a form of unconscious regression to its origins (to its primordial space). In the last moments of the movie, the placenta is identified alongside the Earth, or Gaia, which is presented as a planetary placenta.

The monolith is the symbol of any kind of technology, not just a specific form of technology. As Ricoeur (1965) claimed, each symbol synthetizes two dimensions: the teleological, which looks to the future, and the archaeological, which instead looks to humanity's past. Understanding the symbol means understanding the symbiosis of these two dimensions.

References

Abraham, K. 1924. "The influence of oral eroticism on character-formation." In *Selected Papers of Karl Abraham* (pp. 13–26). New York: Basic Books.

Adams, F., and Aizawa, K. 2008. *The Bounds of Cognition*. New York: Blackwell.

Adams, F., and Aizawa, K. 2010. "Defending the Bounds of Cognition." In R. Menary (ed), *The Extended Mind*. Cambridge, MA: MIT Press.

Akrich, M., and Latour, B. 1992. "A Summary of a Convenient Vocabulary for the Semiotics of Human and Nonhuman Assemblies." In W.E. Bijker and J. Law (eds), *Shaping Technology/Builiding Society*. Cambridge, MA: MIT Press.

Albert, M., and Lee Kleinman, D. 2011. "Bringing Pierre Bourdieu to Science and Technology Studies." *Minerva* 49(3): 263–273.

Ames, M.-G. 2018. "Deconstructing the Algorithmic Sublime." *Big Data and Society* 5: 1–4.

Andersen, P. 1997. *A Theory of Computer Semiotics*. Cambridge: Cambridge University Press.

Anellis, I. 1993. "Review of A Peircean Reduction Thesis." *Modern Logic* 3(4): 401–406.

Angus, I. 2016. *Facing the Anthropocene*. New York: Monthly Press.

Appadurai, A. 1986. "Introduction: Commodities and the Politics of Value." In A. Appadurai (ed), *The Social Life of Things*. Cambridge, MA: Cambridge University Press.

Arendt, H. 2017. *The Origin of Totalitarianism*. New York: Penguin.

Ashby, W.R. 1966. *An Introduction to Cybernetics*. London: Chapman & Hall.

Augustyn, P. 2009. "Uexküll, Peirce, and Other Affinities Between Biosemiotics and Biolinguistics." *Biosemiotics* 2(1): 1–17.

Auroux, S. 1993. *La revolution technologique de la grammatisation*. Paris: Mardaga.

Aydin, C. 2021. *Extimate Technology: Self-formation in a Technological World*. New York: Taylor & Francis.

Bainbridge, W.A., Hart, J., Kim, E.S., and Scassellati, B. 2008. "The Effect of Presence on Human-robot Interaction." In *RO-MAN 2008-The 17th IEEE International Symposium on Robot and Human Interactive Communication* (pp. 701–706). New York: IEEE.

Balazka, D., and D. Rodighiero. 2020. "Big Data and the Little Big Bang: An Epistemological Revolution." *Frontiers in Big Data* 3(31).

Baldini, F. 1998. "Freud's Line of Reasoning. A Note about Epistemic and Clinical Inconsistency of Grünbaum's Argument Pretending to Confute Freud's Therapeutic Approach, with Reference to the Thesis of Stengers on Psychoanalysis." *Psychoanalytische Perspectieven* 32(33): 9–39.

Baldini, F. 2019. "Su alcuni passi cruciali dei testi di Freud e sul loro completo fraintendimento da parte di Lacan." *Metapsychologica. Rivista freudiana di psicanalisi freudiana* 1(1): 13–35.
Baldini, F. 2020. "Nuove considerazioni sul metodo psicanalitico freudiano e in generale sull'architettura empirico-razionale della metapsicologia." *Metapsychologica. Rivista freudiana di psicanalisi freudiana* 2(1): 5–38.
Baldini, F. 2021. "Intervista sulla concezione freudiana della psicanalisi." *Metapsychologica. Rivista freudiana di psicanalisi freudiana* 3(1): 5–43.
Baranzoni, S., Lucci, A., and P. Vignola. 2016. "L'Antropocene: fine, medium o sintomo dell'uomo?" *Lo Sguardo* 3(22): 5–9.
Barbieri, M. (ed). 2008. *Introduction to Biosemiosis*. Dordrecht: Springer.
Barbieri, M. 2012. "Codepoiesis—The Deep Logic of Life." *Biosemiotics* 5(3): 297–299.
Barbieri, M. 2015. *Code Biology*. Dordrecht: Springer.
Barthélémy, J.-H. 2005. *Penser l'individualisme. Simondon et la philosophie de la nature*. Paris: L'Harmattan.
Baker, M. 2016. "1500 Scientists Lift the Lid on Reproducibility." *Nature* 533(7604): 452–454.
Bateson, G. 1972. *Steps to an Ecology of Mind*. London: Aronson.
Baumann, Z. 2000. *Liquid Modernity*. Cambridge: Polity Press.
Becker, J. 2021. "Anthropology, AI, and Robotics." In A. Elliott (ed), *The Routledge Social Sciences Handbook of AI* (pp. 67–89). London: Routledge.
Benedetti, F. 2008. *Placebo Effects: Understanding the Mechanisms in Health and Disease*. Oxford: Oxford University Press.
Benedetti, F. 2015. *Placebo e nocebo, dalla fisiologia alla clinica*. Roma: Giovanni Fioriti.
Bentley-Condit, V.K., and Smith, E.O. 2010. "Animal Tool Use: Current Definitions and an Updated Comprehensive Catalog." *Behaviour* 147(2): 185–221.
Bergson, H. 1965. *Matière et mémoire*. Paris: Puf.
Bertalanffy, L. 1969. *General System Theory*. New York: Braziller.
Bion, W. 1961. *Experiences in Groups and Other Papers*. New York: Routledge.
Bird, J., and Green, D. 2020. "Capitalist Realism and Its Psycho-social Dimensions." *Psychoanalysis, Culture & Society*. doi: 10.1057/s41282-020-00162-9
Bisconti, P. 2021a. PhD Dissertation. (Scuola Superiore Sant'Anna).
Bisconti, P. 2021b. "How Robots' Unintentional Metacommunication Affects Human—Robot Interactions. A Systemic Approach." *Minds and Machines* 31(4): 487–504.
Blease, C., and Kirsh, I. 2016. "The Placebo Effect and Psychotherapy: Implications for Theory, Research and Practice." *Psychology of Consciousness: Theory, Research and Practice* 16(3): 252–260.
Bloor, D. 1976. *Knowledge and Social Imagery*. Chicago, IL: The University of Chicago Press.
Blok, A., Farías, I., and Roberts, C. 2020. *The Routledge Companion to Actor-Network Theory*. London-New York: Routledge.
Bollas, C. 1989. *The Shadow of the Object. Psychoanalysis of the Unthought Known*. London: Routledge.
Boolos, G., Burgess, J., and Jeffrey, R. 2007. *Computability and Logic*. Cambridge: Cambridge University Press.
Borgmann, A. 1984. *Technology and the Character of Contemporary Life: A Philosophical Inquiry*. Chicago, IL: The University of Chicago Press.

Bory, P. 2019. "Deep New: The Shifting Narratives of Artificial Intelligence from Deep Blue to AlphaGo." *Convergence* 25(4): 627–642.

Bostrom, N. 2014. *Superintelligence: Paths, Dangers, Strategies*. Oxford: Oxford University Press.

Bourdieu, P. 1980. *Le sens pratique*. Paris: Minuit.

Bourdieu, P. 1982. *Ce que parler veut dire*. Paris: Fayard.

Bourdieu, P. 1997. *Méditations pascaliennes*. Paris: Seuil.

Bourne, C. 2019. "AI Cheerleaders: Public Relations, Neoliberalism and Artificial Intelligence." *Public Relations Inquiry* 8(2): 109–125.

Braidotti, R. 2013. *The Posthuman*. Cambridge: Polity Press.

Bratton, B. 2015. *The Stack. On Software and Sovereignty*. Cambridge, MA: MIT Press.

Bratton, B. 2021. "Planetary Sapience." *Noema*. www.noemamag.com/planetary-sapience/

Brentari, C. 2016. *Jakob von Uexküll. The Discovery of the Umwelt between Biosemiotics and Theoretical Biology*. Dordrecht: Springer.

Brook, A. 2003. "Kant and Freud." In C. Feltman and M. Cheung Chung (eds), *Psychoanalytic Knowledge* (pp. 1–39). London-New York: Palgrave MacMillian.

Buchli, V. 1999. *An Archeology of Socialism*. Oxford-New York: Berg.

Burch, R. 1993. "A Peircean Reduction Thesis." *Transactions of the Charles S. Peirce Society* 29 (1): 101–107.

Cagna, P. 2019. "Teoria del placebo in medicina e psicologia versus teoria della suggestione in psicoanalisi: una valutazione epistemologica." *Metapsychologica. Rivista freudiana di psicanalisi freudiana* 1(1): 131–145.

Calarco, M. 1972. *Zoographies. The Question of the Animal from Heidegger to Derrida*. New York: Columbia University Press.

Callon, M. 1986. "Éléments pour une sociologie de la traduction. La domestication des coquilles Saint-Jacques dans la Baie de Saint-Brieuc" *L'Année sociologique* 36: 169–208.

Callon, M. (ed). 1989. *La science et ses reseaux*. Paris: La Découverte.

Canguilhem, G. 1965. *La connaissance de la vie*. Paris: Vrin.

Canguilhem, G. 1966. *Le normal et le pathologique*. Paris: Puf.

Canguilhem, G. 2011. *Oeuvres Completes*. Vol. 1. Paris: Vrin.

Canguilhem, G. 2015. *Oeuvres Completes*. Vol. 4. Paris: Vrin.

Carbonell, J.G., Michalski, R.S., Mitchell, T.M. 1983. "An Overview of Machine Learning." In: R.S. Michalski, J.G. Carbonell and T.M. Mitchell (eds), *Machine Learning. Symbolic Computation*. Berlin: Springer. https://doi.org/10.1007/978-3-662-12405-5_1

Carlson, J., and Murphy, R.R. 2005. "How UGVs Physically Fail in the Field." *IEEE Transactions on Robotics* 21: 423–437. doi: 10.1109/TRO.2004.838027

Cavell, M. 1993. *The Psychoanalytic Mind: From Freud to philosophy*. Cambridge, MA: Harvard University Press.

Cellucci, C. 2019. "Diagrams in Mathematics." *Foundations of Sciences* 24: 583–604.

Ceschi, M.V. 2019. "La validità epistemica del metodo d'indagine freudiano: il caso del sogno" *Metapsychologica. Rivista freudiana di psicanalisi freudiana* 1(1): 111–130.

Ceschi, M.V. 2020. "Riflessioni epistemologiche su alcuni aspetti del metodo freudiano" *Metapsychologica. Rivista freudiana di psicanalisi freudiana* 2(1): 39–73.

Ceschi, M.V. 2021. "I limiti metodologici e teorici della ricerca contemporanea in psicoterapia." *Metapsychologica. Rivista freudiana di psicanalisi freudiana* 3(1): 43–63.

Chalmers, D. 2011. "A Computational Foundation for the Study of Cognition." *Journal of Cognitive Science* 12(4): 323–357.

Christin, A. 2018. "Algorithms in Practice: Comparing Web Journalism and Criminal Justice." *Big Data and Society* 5: 1–4.

Christian, B. 2020. *The Alignment Problem*. New York: Norton & Company.

Chun, W. 2011. *Programmed Visions*. Cambridge, MA: MIT Press.

Cimatti, F. 2018. *A Biosemiotics Ontology. The Philosophy of Giorgio Prodi*. Dordrecht: Springer.

Clarizio, E. 2021. *La vie technique. Une philosophie biologique de la technique*. Paris: Hermann.

Coeckelbergh, M. 2011. "Humans, Animals, and Robots: A Phenomenological Approach to Human-Robot Relations." *International Journal of Social Robotics* 3(2): 197–204.

Coeckelberg, M. 2020a. *AI Ethics*. Cambridge, MA: The MIT Press.

Coeckelberg, M. 2020b. "Artificial Intelligence, Responsibility Attribution, and a Relational Justification of Explainability." *Science and Engineering Ethics* 26: 2051–2068.

Coeckelberg, M., and Reijers, W. 2020. *Narrative and Technology Ethics*. New York-London: Palgrave Macmillan.

Coeckelberg, M., Romele, A., and Reijers, W. 2021. *Interpreting Technology: Ricoeur on Questions Concerning Ethics and Philosophy of Technology*. London-New York: Rowman & Littlefield.

Collingridge, D. 1980. *The Social Control of Technology*. New York: St. Martin's Press.

Combes, M. 2012. *Gilbert Simondon and the Philosophy of the Transindividual*. Cambridge, MA: MIT Press.

Copeland, J. 2017. "Turing's Great Invention: The Universal Computing Machine." In J. Copeland, J. Bowen, M. Sprevak and R. Wilson (eds), *The Turing Guide*. Oxford: Oxford University Press.

Copeland, J., and Shagrir, O. 2011. "Do Accelerating Turing Machines Compute the Incomputable?" *Minds and Machines* 21: 221–239.

Cornell, W. 2019. *Self-examination in Psychoanalysis and Psychotherapy*. London: Routledge.

Crawford, K. 2021. *The Atlas of AI. Power, Politics, and the Planetary Costs of Artificial Intelligence*. London: Yale University Press.

Cremerius, J. 1981. *Psicosomatica Clinica*. Roma: Borla.

Crutzen, P. 2002. "Geology of Mankind." *Nature* 413: 23.

Crutzen, P. 2004. "'Anti-Gaia', Box 2.7." In W. Steffen et al. (eds), *Global Change and the Earth System*. Berlin: Springer.

Dalto, S. 2021. "Verità e pragmatica della verità." *Metapsychologica. Rivista freudiana di psicanalisi freudiana* 3(1): 63–95.

Deacon, T. 1997. *The Symbolic Species*. London-New York: Norton & Company.

Deacon, T. 2003. "Universal Grammar and Semiotic Constraints." *Studies in the Evolution of Language* 3: 111–139.

Deacon, T. 2011. *Incomplete Nature*. London-New York: Norton & Company.

De Mijolla, A. (ed). 2002. *The International Dictionary of Psychoanalysis*. New York: Thomson.

De Mul, J. 2021. "The Living Sign. Reading Noble from a Biosemiotic Perspective." *Biosemiotics* https://doi.org/10.1007/s12304-021-09426-y

Derrida, J. 1967. *De la grammatologie*. Paris: Seuil.

Derrida, J. 1972. *La dissémination*. Paris: Seuil.

Derrida, J. 1974. *Of Grammatology*. Baltimore and London: The John Hopkins University Press.

Derrida, J. 1982. *Margins of Philosophy*. Chicago, IL: University of Chicago Press.

Derrida, J. 1993. *Spectres de Marx*. Paris: Galilée.

Derrida, J. 2005. *On Touching—Jean-Luc Nancy*. Stanford: Stanford University Press.

Derrida, J. 2008. *The animal that therefore I am*. New York: Fordham University Press.

Descola, P. 2013. *Beyond Nature and Culture*. Chicago, IL: The University of Chicago Press.

Di Bella, S. 2019. Book Review: *The Age of Surveillance Capitalism* by S. Zuboff. *LSE Review of Books* https://blogs.lse.ac.uk/lsereviewofbooks/

Dignum, V. 2019. *Responsible Artificial Intelligence*. Dordrecht: Springer.

Doyle, B. 2016. "Five Reasons Why Google Glass was a Miserable Failure." *Business2community*. 28 February 2016, https://www.business2community.com/tech-gadgets/5-reasons-google-glass-miserable-failure-01462398

Duffy, B.R. 2003. "Anthropomorphism and the Social Robot." *Robotics and Autonomous Systems* 42(3–4): 177–190. https://doi.org/https://doi.org/10.1016/S0921-8890(02)00374-3

Eco, U. 1975. *Trattato di semiotica generale*. Milano: Bompiani.

Eco, U. 2020. *Lector in Fabula*. Milano: La Nave di Teseo.

Edelman, G. 1987. *Neural Darwinism. The Theory of Neuronal Group Selection*. New York: Basic Books.

Ellen, R. 1988. "Fetishism." *Man* (N.S.) 23: 213–235.

Ellenberger, H. 1970. *The Discovery of the Unconscious*. New York: Basic Books.

Elliott, A. 2014. *Concepts of the Self*. London: Polity Press.

Elliott, A. 2015. *Identity Troubles: An Introduction*. London: Routledge.

Elliott, A. 2018. *The Culture of AI: Everyday Life and the Digital Revolution*. London-New York: Routledge.

Elliott, A. 2021. "The Complex Systems of AI: Recent Trajectories of Social Sciences." In A. Elliott (ed), *The Routledge Social Sciences Handbook of AI* (pp. 13–26). London: Routledge.

Emmeche, C. 1992. "Modeling Life: A Note on the Semiotics of Emergence and Computation in Artificial and Natural Living Systems." In T.A. Sebeok and J. Umiker-Sebeok (eds), *Biosemiotics: The Semiotic Web 1991*. Berlin: Mouton de Gruyter.

Esposito, R. 2022. *Immunità comune. Biopolitica all'epoca della pandemia*. Torino: Einaudi.

Etchegoyen, R.H. 1991. *The Fundamentals of Psychoanalytic Technique*. London: Karnac Books.
Evers, A.W.M., Colloca, L., Blease, C., Annoni, M., Atlas, L.Y., Benedetti, F., Bingel, U., Büchel, C., Carvalho, C., Colagiuri, B., Crum, A.J., Enck, P., Gaab, J., Geers, A.L., Howick, J., Jensen, K.B., Kirsch, I., Meissner, K., Napadow, V., Peerdeman, K.J., Raz, A., Rief, W., Vase, L., Wager, T.D., Wampold, B.E., Weimer, K., Wiech, K., Kaptchuk, T.J., Klinger, R., and Kelley, J.M. 2018. "Implications of Placebo and Nocebo Effects for Clinical Practice: Expert Consensus." *Psychotherapy and Psychosomatics* 87(4): 204–210.
Fadda, E. 2013. *Peirce*. Roma: Carocci.
Farina, A. 2014. *Introduction to Ecological Codes*. www.codebiology.org/introduction_ecological.html
Farrell, B.A. 1981. *The Standing of Psychoanalysis*. Oxford: Oxford University Press.
Feiten. 2020. "Mind After Uexküll: A Foray Into the Worlds of Ecological Psychologists and Enactivists." *Frontiers in Psychology*. https://doi.org/10.3389/fpsyg.2020.00480
Ferrey, A.E., Burleigh, T.J., and Fenske, M.J. 2015. "Stimulus-category Competition, Inhibition, and Affective Devaluation: A Novel Account of the Uncanny Valley." *Frontiers in Psychology* 6 249: 35–49.
Fink, B. 1999. *A Clinical Introduction to Lacanian Psychoanalysis: Theory and Technique*. Cambridge, MA: Harvard University Press.
Fisher, M. 2009. *Capitalist Realism: Is There No Alternative?* London: Zero Books.
Fisher, M. 2013. *Ghosts of My Life. Writings on Depression, Hauntology and Lost Futures*. London: John Hunt Publishing.
Fisher, M. 2016. *The Weird and the Eerie*. London: Repeater Books.
Ficher, M. 2018. *K-punk: The Collected and Unpublished Writings of Mark Fisher*. London: Repeater.
Floridi, L. 2021. *The Logic of Information*. Oxford: Oxford University Press.
Forrester, J. 1984. *Il linguaggio e le origini della psicoanalisi*. Bologna: Il Mulino.
Frabetti, F. 2014. *Software Theory*. London: Rowman & Littlefield.
Freud, S. 1953. *The Standard Edition of the Complete Psychological Works of Sigmund Freud*. Vol. I. London: Hogarth.
Freud, S. 1955. *The Standard Edition of the Complete Psychological Works of Sigmund Freud*. Vol. XVIII. London: Hogarth.
Freud, S. 1957. *The Standard Edition of the Complete Psychological Works of Sigmund Freud*. Vol. XIV. London: Hogarth.
Freud, S. 1958. *The Standard Edition of the Complete Psychological Works of Sigmund Freud*. Vol. XII. London: Hogarth.
Freud, S. 1960. *The Ego and the Id*. New York-London: Norton & Company.
Freud, S. 2003. *Beyond the Pleasure Principle and Other Writings*. New York: Basic Books.
Fukuyama, F. 2002. *Our Posthuman Future: Consequences of the Biotechnology Revolution*. New York: Farrar, Strauss and Giroux.
Gaab, J., Blease, C., Locher, C., and Gerger, H. 2016. "Go Open: A Plea for Transparency in Psychotherapy." *Psychology of Consciousness: Theory, Research and Practice* 3(2): 175–198.
Geiger, R.S. 2018. "Beyond Opening Up the Black Box." *Big Data and Society* 5: 1–4.

Giannopulu, I., and Watanabe, T. 2015. "Conscious/Unconscious Emotional Dialogues in Typical Children in the Presence of an Interactor Robot." In *2015 24th IEEE International Symposium on Robot and Human Interactive Communication (RO-MAN)* (pp. 264–270). New York: IEE Press.
Giddens, A. 1990. *The Consequences of Modernity*. Cambridge: Polity Press.
Gillespie, T. 2014. "The Relevance of Algorithms." In T. Gillespie, P.J. Boczkowski and K.A. Foot (eds), *Media Technologies: Essays on Communication, Materiality, and Society* (pp. 167–194). Cambridge, MA: MIT Press.
Giuliani, M., Mirnig, N., Stollnberger, G., Stadler, S., Buchner, R., and Tscheligi, M. 2015. "Systematic Analysis of Video Data from Different Human—Robot Interaction Studies: A Categorization of Social Signals During Error Situations." *Frontiers in Psychology* 6: 931. doi: 10.3389/fpsyg.2015.00931
Godfrey-Smith, P., and Sterelny, K. 2016, "Biological Information." In E. N. Zalta (ed), *The Stanford Encyclopedia of Philosophy*. https://plato.stanford.edu/archives/sum2016/entries/information-biological
Goode, L. 2018. "Life, But Not as We Know It: AI and the Popular Imagination." *Culture Unbound: Journal of Current Cultural Research* 10(2): 185–207.
Greimas, A. 1987. *On Meaning. Selected Writings in Semiotic Theory*. Minneapolis, MN: University of Minnesota Press.
Greimas, A., and Courtes, J. (eds). 1979. *Sémiotique: dictionnaire raisonné de la théorie du langage*. Paris: Hachette.
Grotsein, J. 2009. *But at the Same Time and on Another Level*. Vol. 1. London: Karnac Books.
Grünbaum, A. 1984. *The Foundations of Psychoanalysis*. Berkeley, CA: University of California Press.
Grunwald, A. 1999. "Technology Assessment or Ethics of Technology?" *Ethical perspectives* 6(2): 170–182.
Grunwald, A. 2009. "Technology Assessment: Concepts and Methods." In A Meijers (ed), *Philosophy of Technology and Engineering Sciences* (pp. 1103–1146). Amsterdam, The Netherlands: North Holland.
Grunwald, A. 2018. *Technology Assessment in Practice and Theory* (1st ed.). London: Routledge.
Guariento, T. 2016. "La disarmonia del mondo. L'Antropocene e l'immagine premoderna della natura." *Lo Sguardo* 3(22): 13–32.
Guma, F. 2019. "L'architettura trascendentale della metapsicologia freudiana (parte prima)." *Metapsychologica. Rivista freudiana di psicanalisi freudiana* 1(1): 51–81.
Guma, F. 2020. "L'architettura trascendentale della metapsicologia freudiana (parte seconda)." *Metapsychologica. Rivista freudiana di psicanalisi freudiana* 2(1): 133–173.
Hägglund, M. 2008. *Radical Atheism. Derrida and the Time of Life*. Stanford: Stanford University Press.
Hägglund, M. 2011. "The Arche-materiality of Time: Deconstruction, Evolution and Speculative Materialism." In J. Elliott and D. Attridge (eds), *Theory after 'Theory'* (pp. 265–277). London: Routledge.
Hamilton, C. 2016. "The Anthropocene as Rupture." *The Anthropocene Review*. doi: 10.1177/2053019616634741
Haraway, D. 1991. *Simians, Cyborgs, and Women: The Reinvention of Nature*. New York: Routledge.

Harman, G. 2009. *Prince of Networks. Bruno Latour and Metaphysics*. Melbourne: re.press.
Hasse, C. 2015. "Multistable Roboethics." In L. Botin, A. Forss, M. Funk, C. Hasse, S.O. Irwin, R. Lally and G. Wellner (eds), *Technoscience and Postphenomenology: The Manhattan Papers* (pp. 169–188). London: Lexington Books.
Hayles, K. 2008. *How We Became Posthuman: Virtual Bodies in Cybernetics, Literature and Informatics*. Chicago, IL: University of Chicago Press.
Héran, F. 1987. "La seconde nature de l'habitus. Tradition philosophique et sens commun dans le langage sociologique." *Revue française de sociologie* 28(3): 385–416.
Hereth Correia, J., Dau, F. 2006. "Two Instances of Peirce's Reduction Thesis." In R. Missaoui and J. Schmidt (eds), *Formal Concepts Analysis* (4th International Conference, ICFCA 2006, Dresden, Germany, February 13–17, 2006. Proceedings). Dordrecht: Springer.
Hoffmeyer, J. 1996. *Signs of Meaning in the Universe*. Bloomington, IN: Indiana University Press.
Honig, S., and Oron-Gilad, T. 2018. "Understanding and Resolving Failures in Human-robot Interaction: Literature Review and Model Development." *Frontiers in Psychology* 9: 861.
Horstmann, A.C., Bock, N., Linhuber, E., Szczuka, J.M., Straßmann, C., and Krämer, N.C. 2018. "Do a Robot's Social Skills and Its Objection Discourage Interactants from Switching the Robot off?" *PLoS One* 13(7). https://doi.org/10.1371/journal.pone.0201581
Hsu, E. 2016. "Accelerated Identity. Five Theses on the Self." In A. Elliott (ed), *The Routledge Social Sciences Handbook of AI* (pp. 24–56). London: Routledge.
Hui, Y. 2016. *On the existence of digital objects*. Minneapolis-London: University of Minnesota Press.
Husserl, E. 1991. *On the Phenomenology of the Consciousness of Internal Time (1893–1917)*. Berlin: Springer.
Husserl, E. 2012. *Logical Investigations Volume 1*. London-New York: Routledge.
Iacovides, I., Cox, A.L., McAndrew, P., Aczel, J., and Scanlon, E. 2015. "Game-play Breakdowns and Breakthroughs: Exploring the Relationship between Action, Understanding, and Involvement." *Human-Computer Interaction* 30(3–4): 202–231.
Ihde, D. 1979. "Heidegger's Philosophy of Technology." In D. Ihde (ed), *Technics and Praxis* (pp. 103–129). Dordrecht: Springer.
Ihde, D. 1990. *Technology and the Lifeworld: From Garden to Earth*. Bloomington, IN: The Indiana University Press.
Introna, L.D. 2015. "Algorithms, Governance, and Governmentality: On Governing Academic Writing." *Science, Technology, & Human Values* 41(1): 17–49.
Johanssen, J. 2019. *Psychoanalysis and Digital Culture*. London-New York: Routledge.
Johanssen, J., and Wang, X. 2021. "Artificial Intuition in Tech Journalism on AI: Imagining the Human Subject." *Human-Machine Communication* 2.
Kant, I. 1995. *Opus Postumum*. Cambridge: Cambridge University Press.
Kelleher, J., and Tierney, B. 2018. *Data Science*. Cambridge, MA: MIT Press.

Kiesler, S., Powers, A., Fussell, S.R., and Torrey, C. 2008. "Anthropomorphic Interactions with a Robot and Robot—like Agent." *Social Cognition* 26(2): 169–181. https://doi.org/10.1521/soco.2008.26.2.169

Kirsh, I., Wampold, B., and Kelley, J.M. 2016. "Controlling for the Placebo Effect in Psychotherapy: Nobel Quest or Tilting at Windmills?" *Psychology of Consciousness: Theory, Research and Practice* 3(2): 121–131.

Klein, M. 1946. "Notes on Some Schizoid Mechanisms." *International Journal of Psychoanalysis* 27: 99–110.

Knafo and LoBosco, eds. 2017. *The Age of Perversion. Desire and Technology in Psychoanalysis and Culture*. Abingdon: Taylor & Francis.

Knappett, C. 2002. "Photographs, Skeuomorphs and Marionettes: Some Thoughts on Mind, Agency and Object." *Journal of Material Culture* 7(1): 97–117

Kohn, E. 2013. *How Forests Think. Toward an Anthropology Beyond the Human*. Berkeley, CA: University of California Press.

Koselleck, R. 2002. *The Practice of Conceptual History*. Princeton, NJ: Princeton University Press.

Krämer, N.C., Eimler, S., Von Der Pütten, A., and Payr, S. 2011. "Theory of Companions: What Can Theoretical Models Contribute to Applications and Understanding of Human-robot Interaction?" *Applied Artificial Intelligence* 25(6): 474–502.

Krämer, S. 2014. "Mathematizing Power, Formalization, and the Diagrammatical Mind, or: What Does Computation Mean?" *Philosophy and Technology* 27(3): 345–357.

Kudina, O., and Verbeek, P.-P. 2019. "Ethics from within: Google Glass, the Collingridge Dilemma, and the Mediated Value of Privacy." *Science, Technology & Human Values*. https://doi.org/10.1177/0162243918793711

Kuhn, T. 1962. *The Structure of Scientific Revolution*. Chicago, IL: The University of Chicago Press.

Kull, K., Deacon, T., Emmeche, C., Hoffmeyer, J., and Stjernfelt, F. 2009. "Theses on Biosemiotics. Prolegomena to a Theoretical Biology." *Biological Theory* 4(2): 167–173.

Kuper, A., and Stone, A. 1982. "The Dream of Irma's Injection: A Structural Analysis." *The American Journal of Psychiatry* 139(10): 1225–1234.

Kurzweil, R. 2005. *The Singularity is Near: When Humans Transcend Biology*. New York: Penguin.

Lacan, J. 2005. *Ecrits: A Selection*. London: Routledge.

Lagoze, C. 2014. "Big Data, Data Integrity, and the Fracturing of the Control Zone." *Big Data and Society* 1: 1–11.

Lakoff, G., and Johnsen, M. 2003. *Metaphors We Live By*. Chicago, IL: The University of Chicago Press.

Lambert, M.J., and Bergin, A.E. 1992. "Achievements and Limitations of Psychotherapy Research." In D.K. Freedheim, H.J. Freudenberger, J.W. Kessler, S.B. Messer, D.R. Peterson, H.H. Strupp and P.L. Wachtel (eds), *History of Psychotherapy: A Century of Change* (pp. 360–390). Washington, DC: American Psychological Association. https://doi.org/10.1037/10110-010

Lami, G. 2019. "Dalla formalizzazione della metapsicologia alla naturalizzazione della matematica." *Metapsychologica. Rivista freudiana di psicanalisi freudiana* 1(1): 81–109.

Lami, G. 2020. "Primi approcci a una dinamica formale della mente." *Metapsychologica. Rivista freudiana di psicanalisi freudiana* 2(1): 219–249.

Landowski, E., and Marrone, G. (eds). 2007. *La società degli oggetti. Problemi di intersoggettività*. Milano: Booklet.

Langley, P. 2011. "The changing science of machine learning." *Machine Learning* 82: 275–279. https://doi.org/10.1007/s10994-011-5242-y

Langs, R. 1984. "Irma's dream and the origins of psychoanalysis" *Psychoanalytic Review* 71(4): 591–617.

Laplanche, J., and Pontalis, J.-B. 1988. *The Language of Psychoanalysis*. London: Routledge.

Laprie, J.-C. 1995. "Dependable Computing and Fault Tolerance: Concepts and Terminology." In *25th International Symposium on Fault-Tolerant Computing, "Highlights from Twenty-Five Years"* (pp. 2–11). Pasadena, CA: IEEE Press.

Latour, B. 1988. *The Pasteurization of France*. Cambridge, MA: Harvard University Press.

Latour, B. 1993. *We Have Never Been Modern*. Cambridge, MA: Harvard University Press.

Latour, B. 1993a. "Portrait de Gaston Lagaffe en philosophe des techniques". In B. Latour (ed), *La clef de Berlin et autres leçons d'un amateur de sciences*. Paris: La Découverte.

Latour, B. 1993b. "La clef de Berlin." In B. Latour (ed), *Petites leçons de sociologie des sciences*. Paris: La Découverte.

Latour, B. 1993c. "Le groom est en grève. Pour l'amour de Dieu, fermez la porte." In B. Latour (ed), *Petites leçons de sociologie des sciences*. Paris: La Découverte.

Latour, B. 1994. "On Technical Mediation." *Common Knowledge* 3/2: 29–64.

Latour, B. 1995. "A Door Must Be Either Open or Shut: A Little Philosophy of Techniques." In A. Feenberg and A. Hannaway (eds), *Technology, and the Politics of Knowledge*. Minneapolis, MN: Indiana University Press.

Latour, B. 1999. *Pandora's Hope. Essay on the Reality of Science Studies*. Cambridge, MA: Harvard University Press.

Latour, B. 2007. *Reassembling the Social*. Oxford: Oxford University Press.

Latour, B. 2008. "A Cautious Prometheus? A Few Steps Toward a Philosophy of Design." In F. Hackne, J. Glynne and V. Minto (eds), *Proceedings of the 2008 Annual International Conference of the Design History Society* (pp. 2–10). Falmouth: Universal Publishers.

Latour, B. 2014. "Comment on Kohn, Eduardo. 2013. *How Forests Think: Toward an Anthropology Beyond the Human*. Berkeley, CA: University of California Press." *Hau: Journal of Ethnographic Theory* 4(2): 261–266.

Latour, B. 2017. *Facing Gaia*. Cambridge: Polity Press.

Latour, B., and Callon, M. 2013. *La science telle qu'elle de fait*. Paris: La Découverte.

Latour, B., and Strum, S.C. 1986. "Human Social Origins: Oh Please, Tell Us Another Story." *Journal of Social and Biological Structures* 9(2): 169–187.

Latour, B., and Strum, S.S. 1987. "Redefining the Social Link: From Baboons to Humans." *Social Science Information* 26(4): 783–802. doi: 10.1177/053901887026004004

Latour, B., and Woolgar, S. 1979. *Laboratory Life. The Construction of Scientific Facts*. Thousand Oaks, CA: Sage Publications.

Lee, M.K. 2018. "Understanding Perception of Algorithmic Decisions: Fairness, Trust, and Emotion in Response to Algorithmic Management." *Big Data and Society* 5: 1–11.

Lenoir, T. 1994. "Was That Last Turn a Right Turn?" *Configurations* 2: 119–136.

Leroi-Gourhan, A. 1943. *L'homme et la matière: évolution et technique*. Paris: Albin Michel.

Leroi-Gourhan, A. 1945. *Milieu et technique*. Paris: Albin Michel.

Leroi-Gourhan, A. 1964. *Le geste et la parole*. Vol. 2. Paris: Albin Michel.

Lovelock, J. 2000. *Gaia. A New Look at the Life on Earth*. Oxford: Oxford University Press.

Lovelock, J. 2019. *Novacene. The Coming Age of Hyperintelligence*. London: Penguin.

Lovelock, J., and L. Margulis. 1974. "Atmospheric Homeostasis by and for the Biosphere: The Gaia Hypothesis." *Tellus* 26(1–2): 2–10. doi: 10.3402/tellusa.v26i1-2.9731

Luborsky, L., McLellan, A.T., Diguer, L., Woody, G., and Seligman, D.A. 1997. "The Psychotherapist Matters: Comparison of Outcomes Across Twenty-two Therapists and Seven Patient Samples." *Clinical Psychology: Science and Practice* 4(1): 53–65.

Luria, M., Reig, S., Tan, X.Z., Steinfeld, A., Forlizzi, J., and Zimmerman, J. 2019. "Re-Embodiment and Co-Embodiment." *Proceedings of the 2019 on Designing Interactive Systems Conference* 633–644. https://doi.org/10.1145/3322276.3322340

Lyotar, J.-F. 1979. *La Condition Postmoderne*. Paris: De Minuit.

MacDorman, K.F., and Ishiguro, H. 2006. "The Uncanny Advantage of Using Androids in Cognitive and Social Science Research." *Interaction Studies* 7(3): 297–337.

MacDorman, K.F., Green, R.D., Ho, C.C., and Koch, C.T. 2009. "Too Real for Comfort? Uncanny Responses to Computer Generated Faces." *Computers in Human Behavior* 25(3): 695–710.

Macho, T. 2011. *Vorbilder*. München: Fink.

Malafouris, L. 2010. "The Brain—Artefact Interface (BAI): A Challenge for Archaeology and Cultural Neuroscience." *Social Cognitive and Affective Neuroscience* 5(2–3): 264–273.

Malafouris, L. 2013. *How Things Shape the Mind. A Theory of Material Engagement*. Cambridge, MA: MIT Press.

Marino, M. 2020. *Critical Code Studies*. Cambridge, MA: The MIT Press.

Marrone, G., Mangano, D. 2018. *Semiotics of Animals in Culture. Zoosemiotics 2.0*. Dordrecht: Springer.

Martini, G. 2020. "Identità e delirio." In V. Busacchi and G. Martini (eds), *L'identità in questione. Saggio di psicoanalisi ed ermeneutica* (pp. 34–65). Milano: Jaca Book.

Masaaki, K. 2020. "Human-Computer Interaction Multimodal and Natural Interaction PART2." In *22nd International Conference, HCII 2020*. Berlin: Springer.

Massa, N., Bisconti, P., and Nardi, D. 2022. "The Psychological Implications of Companion Robots: A Theoretical Framework and an Experimental Setup." *International Journal of Social Robotics* 2: 1–14.

Merton, R. 1973. *The Sociology of Science. Theoretical and Empirical Investigations*. Chicago, IL: The University of Chicago Press.

Meunier, J-G. 2021. *Computational Semiotics*. London: Bloomsbury.
Millar, I. 2021. *The Psychoanalysis of Artificial Intelligence*. London-New York: Palgrave Macmillan.
Mills, J. 2000. "Dialectical Psychoanalysis: Toward Process Psychology." *Psychoanalysis and Contemporary Thought* 23(3): 417–450.
Mitchell, S.A., and M.J. Black. 1995. *Freud and Beyond: A History of Modern Psychoanalytic Thought*. New York: Basic Books.
Miyazaki, S. 2016. "Algorhythmic Ecosystems: Neoliberal Couplings and their Pathogenesis 1960—Present." In R. Seyfert and J. Roberge (eds), *Algorithmic Cultures* (pp. 20–32). London: Routledge.
Mori, M., MacDorman, K.F., and Kageki, N. 2012. "The Uncanny Valley." *IEEE Robotics & Automation Magazine* 19(2): 98–100.
Morozov, E. 2013. *To Save Everything, Click Here. Technology, Solutionism and the Urge to Fix Problems that Don't Exist*. New York: Public Affairs.
Nadin, M. 2007. "Semiotic Machine." *Public Journal of Semiotics* 1(1): 57–75.
Nancy, J.-L. 2010. *L'Intrus*. Paris: Galilée.
Natale, S., and Ballatore, A. 2017. "Imagining the Thinking Machine: Technological Myths and the Rise of Artificial Intelligence." *Convergence* 26(1): 3–18.
Nathan, T., and Zajde, N. 2013. *Psychothérapie démocratique*. Paris: Odile Jacob.
Norman, D. 2004. *Emotional Design*. New York: Basic Books.
Nusselder, A. 2009. *Interface Fantasy. A Lacanian Cyborg Ontology*. Cambridge, MA: MIT Press.
Ogden, C.K., and Richards, I.A. 1923. *The Meaning of Meaning*. New York: Harcourt.
Ogden, T. 1982. *Projective Identification and Psychotherapeutic Technique*. New York: Jason Aronson.
Open Science Collaboration. 2015. "Estimating the Reproducibility of Psychological Science." *Science*. www.science.org/doi/10.1126/science.aac4716
Papilloud, C. 2018. *Sociology Through Relation. Theoretical Assessments from the French Tradition*. London-New York: Palgrave.
Pasquale, F. 2015. *The Black Box Society: The Secret Algorithms that Control Money and Information*. Cambridge, MA: Harvard University Press.
Pavanini, M. 2022. "Multistability and Derrida's Différance: Investigating the Relations between Postphenomenology and Stiegler's General Organology." *Philosophy and Technology* 35(1): 1–22.
Pearl, J., and Mackenzie, D. 2018. *The Book of Why. The New Science of Cause and Effect*. New York: Basic Books.
Peirce, C. 1931–1958. *Collected Papers*. Edited by C. Hartshorn and P. Weiss. Cambridge, MA: Harvard University Press.
Peirce, C. 1976. *The New Elements of Mathematics*. Edited by C. Eisele. The Hague: Mouton Press.
Piccinini, G. 2015. *Physical Computation*. Oxford: Oxford University Press.
Piccinini, G. 2016. "The Computational Theory of Cognition." In V.C. Müller (ed), *Fundamental Issues of Artificial Intelligence*. Dordrecht: Springer.
Plessner, H. 2019. *Levels of Organic Life and the Human*. New York: Fordham University Press.
Popper, K. 2005. *The Logic of Scientific Discovery*. London: Routledge.

Possati, L. 2021. *The Algorithmic Unconscious. How Psychoanalysis Helps in Understanding AI*. London-New York: Routledge.

Possati, L. 2022. *Software as Hermeneutics*. London-New York: Palgrave McMillian.

Possati, L. 2022b. "Psychoanalyzing Artificial Intelligence: The Case of Replika." *AI & Society* 1–14.

Powers, T., and Ganascia, J.-C. 2020. "The Ethics of the Ethics of AI." In M. Dubber, F. Pasquale and S. Das (eds), *The Oxford Handbook of Ethics of AI*. Oxford: Oxford University Press.

Primiero, G. 2020. *On the Foundations of Computing*. Oxford: Oxford University Press.

Prodi, G. 1977. *La basi materiali della significazione*. Milano: Bompiani.

Prodi, G. 1982. *La storia naturale della logica*. Milano: Bompiani.

Prodi, G. 1988. "Material Bases of Signification." *Semiotica* 69(3–4): 191–241.

Prodi, G. 1989. *L'individuo e la sua firma. Biologia e cambiamento antropologico*. Bologna: Il Mulino.

Putnam, H. 1967. *Psychological Predicates*. Pittsburgh, PA: University of Pittsburgh Press.

Rahwan, I., Cebrian, M., Obradovich, O., Bongard, J., Bonnefon, J.-F., Breazeal, C., Crandall, J., Christakis, N., Couzin, I., Jackson, M.O., Jennings, N., Kamar, E., Kloumann, I., Larochelle, H., Lazer, D., McElreath, R., Mislove, A., Parkes, D., Pentland, A., Roberts, M., Shariff, A., Tenenbaum J., and Wellman, M. 2019. "Machine Behavior." *Nature* 568: 477–486.

Ricoeur, P. 1965. *De l'interprétation. Essai sur Freud*. Paris: Seuil.

Ricoeur, P. 1969. *Le conflit des interprétations*. Paris: Seuil.

Ricoeur, P. 1983. *Temps et récit 1*. Paris: Seuil.

Rip, A., Misa, T.J., and Schot, J. (eds). 1995. *Managing Technology in Society*. London: Pinter Publishers.

Roazen, P. 1990. *Brother Animal: The Story of Freud and Tausk*. Piscataway, NJ: Transaction Publishers.

Romele, A. 2019. *Digital Hermeneutics*. London-New York: Routledge.

Romele, A. 2020. "Technological Capital: Bourdieu, Postphenomenology, and the Philosophy of Technology. Beyond the Empirical Turn." *Philosophy & Technology* https://doi.org/10.1007/s13347-020-00398-4

Rosa, H. 2013. *Social Acceleration: A Theory of Modernity*. New York: Columbia University Press.

Rosenberger, R. 2018. "Why it Takes both Postphenomenology and STS to Account for Technological Mediation." In J. Aagaard et al. (eds), *Postphenomenological Methodologies: New Ways in Mediating Techno-Human Relationships*. New York-London: Lexington Books.

Rosenberger, R., and Verbeek, P. 2015. *Postphenomenological Investigations: Essays on Human-technology Relations*. Lanham, MD: Lexington Books.

Rosenthal, D., and Frank, J.D. 1956. "Psychotherapy and the Placebo Effect." *Psychological Bulletin* 53(4): 294–302.

Rosenthal-Von Der Pütten, A.M., Schulte, F.P., Eimler, S.C., Sobieraj, S., Hoffmann, L., Maderwald, S., Brand, M., and Krämer, N.C. 2014. "Investigations on Empathy Towards Humans and Robots using fMRI." *Computers in Human Behavior* 33: 201–212. https://doi.org/10.1016/j.chb.2014.01.004

Ross, R., Collier, R., and O'Hare, G.M. P. 2004. "Demonstrating Social Error Recovery with AgentFactory." In *3rd International Joint Conference on Autonomous Agents and Multi-agent Systems (AAMAS04)* (pp. 1424–1425). New York, NY: IEEE Press.

Russell, S. 2019. *Human Compatible. AI and the Problem of Control.* New York: Penguin.

Russell, S., and Norvig, P. 2016. *Artificial Intelligence. A Modern Approach.* London: Pearson.

Salanskis, M. 2013. *L'herméneutique formelle. L'infini, le continu, l'espace.* Paris: Klincksieck.

Salvador, L. 2020. "Natura e funzioni del linguaggio e del pensiero nella concezione freudiana." *Metapsychologica. Rivista freudiana di psicanalisi freudiana* 2(1): 191–219.

Salvador, L. 2021. "Elementi di critica al concetto di psicoterapia scientifica." *Metapsychologica. Rivista freudiana di psicanalisi freudiana* 3(1): 155–179.

Satake, S., Kanda, T., Glas, D.F., Imai, M., Ishiguro, H., and Hagita, N. 2009. "How to Approach Humans? Strategies for Social Robots to Initiate Interaction." In *Proceedings of the 4th ACM/IEEE International Conference on Human Robot Interaction* (pp. 109–116). New York, NY: IEEE Press.

Schneider, S., Miller, J., Crist, E., and Boston, P. 2004. *Scientists Debate Gaia.* Cambridge, MA: MIT Press.

Seaver, N. 2017. "Algorithms as Culture: Some Tactics for the Ethnography of Algorithmic Systems." *Big Data and Society* 4: 1–12.

Sebeok, T. 2001a. "Biosemiotics: Its Roots, Proliferation, and Prospects." *Semiotica* 134(1/4): 61–78.

Sebeok T. 2001b. "Nonverbal Communication." In P. Cobley (ed), *The Routledge Companion to Semiotics and Linguistics.* London: Routledge.

Sebeok T. 2001c. *Signs: An Introduction to Semiotics.* Toronto: University of Toronto Press.

Seibt, J. 2016. "Integrative Social Robotics, Value-driven Design, and Transdisciplinarity." *Interaction Studies* 21(1): 111–144.

Seibt, J. 2017. "Towards an Ontology of Simulated Social Interaction: Varieties of the "As If" for Robots and Humans." In R. Hakli and J. Seibt (eds), *Sociality and Normativity for Robots* (pp. 11–39). Berlin: Springer.

Seibt, J. 2018. "Classifying Forms and Modes of Co-Working in the Ontology of Asymmetric Social Interactions (OASIS)." In *Robophilosophy/TRANSOR* (pp. 133–146). Amsterdam: IOS Press.

Seibt, J., Vestergaard, C., and Damholdt, M.F. 2020. "Sociomorphing, Not Anthropomorphizing: Towards a Typology of Experienced Sociality." *Culturally Sustainable Social Robotics--Proceedings of Robophilosophy* 51–67.

Sennett, R. 1998. *The Corrosion of Character, The Personal Consequences of Work In the New Capitalism.* New York-London: Norton.

Serholt, S. 2018. "Breakdowns in Children's Interactions with a Robotic Tutor: A Longitudinal Study." *Computers in Human Behavior* 81: 250–264.

Serres, M. 1969. *Hermès I. La communication.* Paris: Minuit.

Serres, M. 1972. *Hermès II. L'interférence.* Paris: Minuit.

Serres, M. 1974. *Hermès III. La traduction.* Paris: Minuit.

Serres, M. 1977. *Hermès IV. La distribution.* Paris: Minuit.

Serres, M. 1980. *Hermès V. Le passage du Nord-Ouest*. Paris: Minuit.

Serres, M. 1995. *The Natural Contract*. Ann Arbor, MI: The University of Michigan Press.

Shagrir, O. 2006. "Why We View the Brain as a Compute." *Synthese* 153(3): 393–416.

Shannon, C., and W. Weaver. 1964. *The Mathematical Theory of Communication*. Urbana, IL: University of Illinois Press.

Sharp, H., Rogers, Y., and Preece, J. 2007. *Interaction Design: Beyond Human-computer Interaction* (2nd ed.). New York: John Wiley & Sons.

Shaw, R. 2015. "Big Data and Reality." *Big Data and Society* 2: 1–4.

Shimada, M., Minato, T., Itakura, S., and Ishiguro, H. 2006. "Evaluation of Android Using Unconscious Recognition." In *2006 6th IEEE-RAS International Conference on Humanoid Robots* (pp. 157–162). New York, NY: IEEE Press.

Simondon, G. 1965. "L'imagination et l'invention." *Bulletin de psychologie* 19(246): 395–414.

Simondon, G. 2005. *L'individuation à la lumière des notions de forme et d'information*. Paris: Millon.

Simondon, G. 2014. *Imagination et invention (1965–1966)*. Paris: Puf.

Sloterdijk, P. 2011–2016. *Spheres 1–3*. Cambridge, MA: MIT Press.

Sloterdijk, P. 2013. *In the World Interior of Capital*. Cambridge: Polity Press.

Sloterdijk, P., and Heinrichs, H.J. 2001. *Die Sonne und der Tod*. Frankfurt am Main: Suhrkamp.

Skiena, S. 2008. *The Algorithm Design Manual*. Dordrecht: Springer.

Spinks, C.W. 1991. *Peirce and Triadomania: A Walk in the Semiotic Wilderness*. Berlin-New York: de Gruyter.

Sprengnether, M. 2003. "Mouth to Mouth: Freud, Irma, and the Dream of Psychoanalysis." *American Imago* 60: 259–284.

Stänicke, E., Zachrisson, A., and Vetlesen, A.J. 2020. "The Epistemological Stance of Psychoanalysis: Revisiting the Kantian Legacy." *The Psychoanalytic Quarterly* 89(2): 281–304.

Steinbauer, G. 2013. "A Survey about Faults of Robots Used in RoboCup." In Z.T. van der, X. Chen, P. Stone and L.E. Sucar (eds), *Lecture Notes in Computer Science*. Vol. 7500 (pp. 344–355). Berlin: Springer.

Stengers, I. 2015. *In Catastrophic Times*. Lüneburg: Meson Press.

Sterelny, K. 2004. "Externalism, Epistemic Artefacts and the Extended Mind." In R. Schantz (ed), *The Externalist Challenge: New Studies on Cognition and Intentionality*. Berlin: De Gruyter.

Stern, J. 2010. "Bourdieu, Technique, and Technology." *Cultural Studies* 17(3–49): 367–389.

Stiegler, B. 1998. *Technics and Time 1: The Fault of Epimetheus*. Stanford: Stanford University Press.

Stiegler, B. 2006. "Anamnesis and Hypomnesis." In L. Armand and A. Bradley (eds), *Technicity*. Prague: Literaria Pragensia.

Stiegler, B. 2009. "Teleologics of the Snail: The Errant Self Wired to a WiMax Network." *Theory, Culture & Society* 26(2–3): 33–45.

Stiegler, B. 2014. *Symbolic Misery 1: The Hyperindustrial Epoch*. London: Wiley.

Stiegler, B. 2015. *Symbolic Misery 2: The Katastrophé of the Sensible*. London: Wiley.

Stiegler, B. 2018. "Artificial Stupidity and Artificial Intelligence in the Anthropocene." *Academia. edu*. www. academia.edu/37849763/Bernard_Stiegler_Artificial_Stupidity_and_Artificial_Intelligence_in_the_Anthropocene_2018_.

Stiegler, B. 2020. "Elements for a general organology." *Derrida Today* 13(1): 72–94.

Stjernfelt, F. 2000. "Diagrams as Centerpiece of a Peircean Epistemology." *Transactions of the Charles S. Peirce Society* 36/3: 357–384.

Steffen, W., Sanderson, A., Tyson, P. D., et al. 2005. *Global Change and the Earth System*. Berlin: Springer.

Storni, C. 2015. "Notes on ANT for Designers: Ontological, Methodological and Epistemological Turn in Collaborative Design." *CoDesign* 11(3–4): 166–178.

Taylor, A. 2019. "The Data Center as Technological Wilderness." *Culture Machine*, 18. https://culturemachine.net/vol-18-the-nature-of-data-centers/data-center-as-techno-wilderness/

Tegmark, M. 2017. *Life 3.0. Being Human in the Age of Artificial Intelligence*. New York: Penguin.

Tonnessen, M. 2009. "Umwelt Transitions: Uexküll and Environmental Change" *Biosemiotics* 2: 47–64.

Turing, A. 1950. "Computing Machines and Intelligence." *Mind* 59(236): 433–460.

Turing, A.M. 1936. "On Computable Numbers, with an Application to the Entscheidungsproblem." *Proceedings of the London Mathematical Society* 42: 230–265.

Turkle, S. 1984. *The Second Self: Computer and the Human Spirit*. Cambridge, MA: MIT Press.

Turkle, S. 1988. "Artificial Intelligence and Psychoanalysis: A New Alliance." *Daedalus* 117(1): 241–268.

Turkle, S. 1995. *Life on the Screen: Identity in the Age of the Internet*. New York: Simon & Schuster.

Turkle, S. 2011a. *Alone Together*. New York: Basic Books.

Turkle, S. (ed). 2011b. *Evocative Objects: Things We Think with*. Cambridge, MA: MIT press.

Turner, J. 2019. *Robot Rules: Regulating Artificial Intelligence*. London: Palgrave.

van de Poel, I. 2013. "Translating Values into Design Requirements." In D. Michelfelder, N. McCarthy, D. Goldberg (eds), *Philosophy and Engineering: Reflections on Pratice, Principles, and Process*. Berlin: Springer.

Varela, F., and Maturana, H. 1992. *Tree of Knowledge: The Biological Roots of Human Understanding*. Boulder, CO: Shambhala Publications.

Ventura Bordenca, I. 2021. "Introduzione. Ripensare gli oggetti, riprogettare la società." In B. Latour, *Politiche del design. Semiotica degli artefatti e forme della socialità*. Udine: Mimesis.

Verbeek, P. 2005. "Artifacts and Attachment: A Post-script Philosophy of Mediation." In H. Harbers (ed), *Inside the Politics of Technology: Agency and Normativity in the Co-production of Technology and Society* (pp. 125–146). Amsterdam: Amsterdam University Press.

Verbeek, P. 2008. "Morality in Design: Design Ethics and the Morality of Technological Artifacts". In: *Philosophy and Design*. Springer, Dordrecht.

Verbeek, P. 2011. *Moralizing Technology*. Chicago, IL: University of Chicago Press.

Verbeek, P. 2015. "COVER STORY beyond Interaction: A Short Introduction to Mediation Theory." *Interactions* 22(3): 26–31.

Vial, S. 2019. *Being and the Screen*. Cambridge, MA: The MIT Press.

Vitale, F. 2020. "Making the différance: Between Derrida and Stiegler." *Derrida Today* 13(1): 1–16.

Viveiros de Castro, E. 2009. *Métaphysiques cannibales*. Paris: Puf.

von Foester, H. 2002. *Understanding. Essays in Cybernetics and Cognition*. Berlin: Springer.

von Uexküll, J. 1909. *Umwelt und Innenwelt der Tiere*. Berlin: Springer.

von Uexküll, J. 2010. *A Foray into the Worlds of Animals and Humans, with: A Theory of Meaning*. Minneapolis, MN; London: University of Minnesota Press.

Wampol, B.E., Minami, T., Tierney, S.C., Baskin, T.W., and Biati, K.S. 2005. "The Placebo is Powerful: Estimating Placebo Effects in Medicine and Psychotherapy from Randomized Critical Trials." *Journal of Clinical Psychology* 6(7): 835–854.

Watzlawick, P., Bavelas, J.B., and Jackson, D.D. 2011. *Pragmatics of Human Communication: A Study of Interactional Patterns, Pathologies, and Paradoxes*. New York: Norton & Company.

Wiener, N. 1948. *Cybernetics*. Cambridge, MA: MIT Press.

Williams, I. 2020. "Contemporary Application of Ant: An Introduction." In I. Williams (ed), *Contemporary Applications of Actor Network Theory* (pp. 1–13). Singapore: Palgrave Macmillan.

Winnicott, D.W. 2005. *Playing and Reality*. London and New York: Routledge.

Wooldridge, M. 2021. *The Road to Conscious Machines. The Story of AI*. New York: Penguin.

Wrangham, R. 2009. *Catching Fire: How Cooking Made Us Human*. New York: Basic books.

Yoon, C. 2018. "Assumptions that led to the failure of Google Glass." https://medium.com/nyc-design/the-assumptions-that-led-to-failures-of-google-glass-8b40a07cfa1e

Yuan, L. (Ivy), and Dennis, A.R. 2019. "Acting Like Humans? Anthropomorphism and Consumer's Willingness to Pay in Electronic Commerce." *Journal of Management Information Systems* 36(2): 450–477. https://doi.org/10.1080/07421222.2019.1598691

Zizek, S. 2008. *The Plague of Fantasies*. London-New York: Verso.

Zuboff, S. 2019. *The Age of Surveillance Capitalism. The Fight for a Human Future and the New Frontier of Power*. New York: Public Affairs.

Index

Note: Page numbers in *italics* indicate a figure and page numbers in **bold** indicate a table on the corresponding page.

For Product Safety Concerns and Information please contact our EU
representative GPSR@taylorandfrancis.com
Taylor & Francis Verlag GmbH, Kaufingerstraße 24, 80331 München, Germany

www.ingramcontent.com/pod-product-compliance
Ingram Content Group UK Ltd.
Pitfield, Milton Keynes, MK11 3LW, UK
UKHW020133200926
475702UK00005B/732

* 9 7 8 1 0 3 2 3 8 5 5 1 8 *